PRINCIPLES OF

LIBRARY OF MATHEMATICS

edited by

WALTER LEDERMANN

D.Sc., Ph.D., F.R.S.Ed., Professor of Mathematics, University of Sussex

Linear Equations	P. M. Cohn
Sequences and Series	J. A. Green
Differential Calculus	P. J. Hilton
Elementary Differential Equations and Operators	G. E. H. Reuter
Partial Derivatives	P. J. Hilton
Complex Numbers	W. Ledermann
Principles of Dynamics	M. B. Glauert
Electrical and Mechanical Oscillations	D. S. Jones
Vibrating Strings	D. R. Bland
Vibrating Systems	R. F. Chisnell
Fourier Series	I. N. Sneddon
Solutions of Laplace's Equation	D. R. Bland
Solid Geometry	P. M. Cohn
Numerical Approximation	B. R. Morton
Integral Calculus	W. Ledermann
Sets and Groups	J. A. Green
Differential Geometry	K. L. Wardle
Probability Theory	A. M. Arthurs

PRINCIPLES OF DYNAMICS

BY

M. B. GLAUERT

ROUTLEDGE & KEGAN PAUL
LONDON

First published 1960
by Routledge & Kegan Paul Ltd
Broadway House, 68-74 *Carter Lane*
*London, E.C.*4

Printed in Great Britain by
Latimer, Trend & Co Ltd, Plymouth

Reprinted 1963, 1965 *and* 1969

SBN 7100 4348 1

Preface

THE purpose of this book is to give a systematic account of the basis of classical dynamics, a real appreciation of which is most desirable for students of science and engineering as a background to further studies in their own particular fields. Some of these further developments are treated in other books of this series (e.g. D. S. Jones, *Electrical and Mechanical Oscillations*; R. F. Chisnell, *Vibrating Systems*; D. R. Bland, *Vibrating Strings*).

The first chapter is devoted to a comprehensive account of vector algebra, with illustrations from geometry. The remainder of the book deals with the theory of dynamics in three dimensions, full use being made of vector methods where appropriate. My aim has been to keep the logical thread of the argument well to the fore, so that the reader may see clearly exactly what is being assumed, to what extent new dynamical concepts are deductions from what has gone before, and which principles are universal and which apply only in a limited range of situations.

No previous knowledge of dynamics is assumed, although no doubt most readers will have studied the subject to some extent. I advise such readers to follow the arguments with particular care, since it is often less easy to improve a superficial and possibly inaccurate idea than to grasp a fresh one.

M. B. GLAUERT

The University,
Manchester

Contents

CHAPTER ONE
Vector Algebra

1.1. DEFINITION OF A VECTOR

The value of a physical quantity can often be specified by a single number, giving the magnitude of the quantity in terms of the units we are using. Such a quantity is called a *scalar*, and familiar examples are mass, temperature, electric charge and kinetic energy. But sometimes the quantity has direction as well as magnitude, as in the cases of displacement, velocity and force. These are called *vector* quantities.

Ordinary algebra is quite capable of dealing with the analysis of vector quantities, but the formulae and equations are apt to become lengthy and obscure, and it is much more satisfactory to develop a suitable vector algebra for the purpose. This first chapter will be devoted to a full treatment of the subject. We shall start from a mathematical definition of a vector, and shall proceed to work out the algebraic consequences of this definition with full regard for mathematical exactness, for though some properties of vectors are closely similar to those of scalars, others are strikingly different. To ascertain whether the analysis is applicable to any particular physical quantity it will then merely be necessary to check whether the quantity satisfies our definition. A serious student of physical science will do well to familiarize himself with vector algebra to such an extent that he can think in vectors, so that he can appreciate a vector equation directly without having to translate it back into its equivalent non-vector form. He will find this ability invaluable in gaining a real understanding of the basic principles of many physical subjects, including dynamics.

We shall denote vectors by letters in bold type, e.g. **a**, **b**.

A quantity **a** will be said to be a vector if it obeys the two parts of the following definition.

Condition 1: **a** *has a magnitude and a direction.*

Condition 2: *the sum of* **a** *and* **b** *is a vector* **a**+**b**, *given by the parallelogram law.*

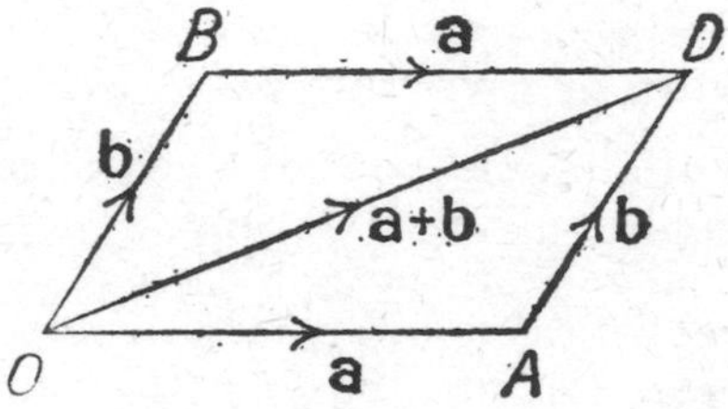

Figure 1

The parallelogram law is illustrated in Fig. 1. Pictorially, the vector **a** is most readily shown by a line of length equal to the magnitude of **a**, in suitable units, and in the same direction as **a**. The parallelogram law may be written as

$$\overrightarrow{OA}+\overrightarrow{OB}=\overrightarrow{OD}$$

where we introduce the notation that $\overrightarrow{OA}$ denotes the vector represented in magnitude and direction by the line OA. Note that $\overrightarrow{OA}$ and $\overrightarrow{BD}$ are both equal to **a**, and $\overrightarrow{OB}$ and $\overrightarrow{AD}$ are both equal to **b**.

The truth of the parallelogram law is evident for several well-known physical quantities of a vector nature. If **a** and **b** are displacements in Euclidean space, successive displacements **a** and **b** produce a displacement such as that from O to D in Fig. 1. If **a** and **b** are forces, the resultant force, as given by the parallelogram of forces, is again as indicated in Fig. 1.

It might be thought that Condition 2 is superfluous. This is not so, for two reasons. Mathematically, without Condition 2 we could not show that **a**+**b** was a vector at all, and so our vector algebra would never get started. And physically, we can find examples in which Condition 1 is obeyed but Condition 2 is not. Consider the displacements **a**, 1000 miles north, and **b**, 1000 miles east, on the surface of the earth. Having magnitude and direction, these satisfy Condition 1. However, the displacement **a** followed by **b** does not bring one to the same point as does **b** followed by **a**, for the longitude change produced by **b** depends upon the latitude. The difference is most clear for a starting point near the south pole. Condition 2 requires $\mathbf{a}+\mathbf{b}=\mathbf{b}+\mathbf{a}$, since the parallelogram of Fig. 1 is the same in each case, and hence displacement over the surface of the earth is not a vector according to our definition. Another example is provided by the rotation of a rigid body pivoted at a fixed point. The angle turned and the axis of rotation specify magnitude and direction, but it can be verified that the order in which two rotations are performed affects the final position of the body. For neither of these examples is vector algebra applicable. We shall call quantities which have magnitude and direction *vector quantities*. Only when we have shown that they obey Condition 2 may we call them *vectors*.

Many authors make a distinction between what they term free vectors and localized vectors, the latter having a specified point or line of application, as when a force is applied at a particular point. Thus in Fig. 1 $\overrightarrow{OA}$ and $\overrightarrow{BD}$ would be different, considered as localized vectors, while according to our definition they are identical. In this book we shall not use the concept of a localized vector. It is true that for the calculation of the moment of a force its line of action must be known; but similarly the location of a mass is required for the calculation of the centre of mass, yet we do not normally refer to mass as a localized scalar or seek to establish the algebra of localized scalars. To us, then, a vector is given by Conditions 1 and 2. Of course this does not in any way imply that the vector giving the magnitude and direction of a physical quantity is the only information we possess or need to possess about the quantity in question.

It may be noted that we have nowhere stated that a vector is in three dimensions. Although in the applications of this book, vectors in two and three dimensions are all that are required, Conditions 1 and 2 are equally suitable for defining vectors in any greater number of dimensions.

1.2. PROPERTIES OF A VECTOR

To discuss vectors we require symbols to denote magnitude and direction separately. We shall write the magnitude or *modulus* of **a** as a, or sometimes $|\mathbf{a}|$, and we shall specify the direction by $\hat{\mathbf{a}}$, a vector in the same direction as **a** and with modulus unity; such a vector is called a *unit vector*.

Without proof, we must not assume that vectors obey the usual laws of algebra. The *commutative law* of addition

$$\mathbf{a}+\mathbf{b}=\mathbf{b}+\mathbf{a}$$

is satisfied since the parallelogram of Fig. 1 is the same in each case. The *associative law* of addition

$$\mathbf{a}+(\mathbf{b}+\mathbf{c})=(\mathbf{a}+\mathbf{b})+\mathbf{c}$$

is also true, as is seen from Fig. 2. From Condition 2 both

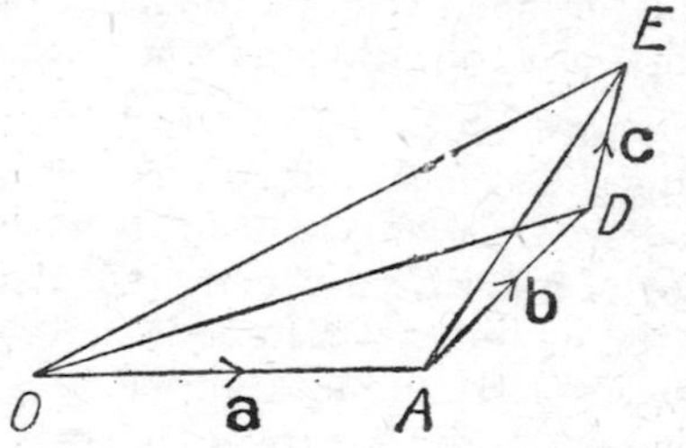

Figure 2

sides of this equation are equal to $\overrightarrow{OE}$, since $\mathbf{b}+\mathbf{c}=\overrightarrow{AE}$ and $\mathbf{a}+\mathbf{b}=\overrightarrow{OD}$. Thus the brackets are unnecessary, $\mathbf{a}+\mathbf{b}+\mathbf{c}$

being uniquely defined. By repeated applications of these laws we see that $\mathbf{a}+\mathbf{b}+\ldots+\mathbf{n}$ is a vector with a unique meaning, not depending on the order in which the individual vectors are written.

The vector $\mathbf{a}+\mathbf{a}$ has modulus $2a$ and direction $\hat{\mathbf{a}}$; we write this vector as $2\mathbf{a}$. Similarly it is natural to write as $k\mathbf{a}$ the vector of modulus ka and direction $\hat{\mathbf{a}}$, where k is any positive number, not necessarily an integer. Then it is easy to see that

$$k\mathbf{a}+l\mathbf{a}=(k+l)\mathbf{a},\quad k(l\mathbf{a})=kl\mathbf{a}=l(k\mathbf{a}),$$

and in particular

$$\mathbf{a}=a\hat{\mathbf{a}}. \tag{1}$$

Also, from Fig. 3,

$$k\mathbf{a}+k\mathbf{b}=k(\mathbf{a}+\mathbf{b}).$$

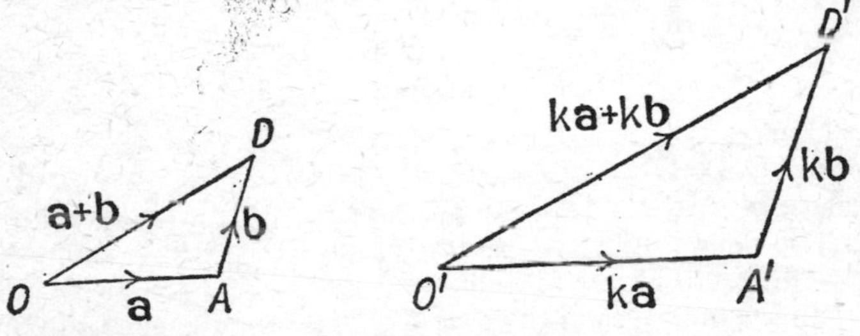

Figure 3

The triangles OAD, $O'A'D'$ are similar and hence $O'D'=kOD$ and $O'D'$ is parallel to OD, i.e. $\overrightarrow{O'D'}=k\overrightarrow{OD}$. Thus the usual laws apply for scalar multiplication.

We have so far required k to be positive. Now if $\mathbf{a}+\mathbf{b}=0$, so that in Fig. 1 the point D coincides with O, it is natural to write $\mathbf{b}=-\mathbf{a}$. The vector $-\mathbf{a}$ has modulus a and direction $-\hat{\mathbf{a}}$, for the modulus of a vector is never negative. If $\overrightarrow{OA}=\mathbf{a}$, then $\overrightarrow{AO}=-\mathbf{a}$. Using this definition, it is easy to see that the general laws proved above are equally true whether k and l are positive or negative.

In three dimensions, we can express any vector **a** as a sum of vectors in three given non-coplanar directions. Consider the parallelepiped shown in Fig. 4, with edges in the given directions and diagonal $\overrightarrow{OA}=\mathbf{a}$, which we can always construct uniquely. By the parallelogram law,

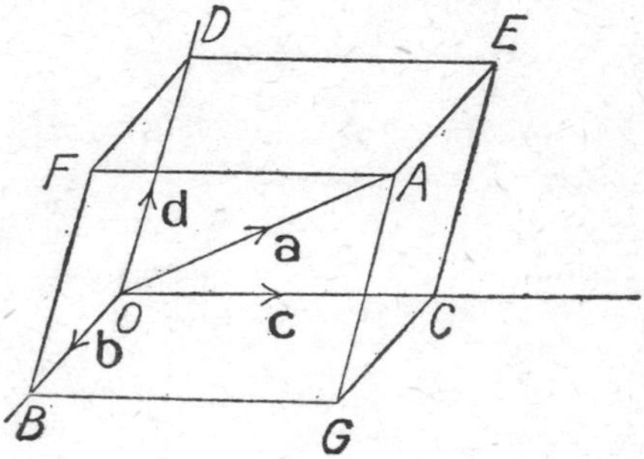

Figure 4

$$\mathbf{a}=\overrightarrow{OA}=\overrightarrow{OB}+\overrightarrow{BG}+\overrightarrow{GA}=\overrightarrow{OB}+\overrightarrow{OC}+\overrightarrow{OD}=\mathbf{b}+\mathbf{c}+\mathbf{d}$$

where **b**, **c**, **d** are in the three given directions.

The most important application of this result is when the given directions are mutually perpendicular, so that **b**, **c**, **d** are in the direction of the axes of the rectangular co-ordinates x, y, z. We introduce unit vectors **i**, **j**, **k** in the co-ordinate directions. Then if a_x, a_y, a_z are the lengths of OB, OC, OD measured in these directions, $\overrightarrow{OB}=a_x\mathbf{i}$, $\overrightarrow{OC}=a_y\mathbf{j}$, $\overrightarrow{OD}=a_z\mathbf{k}$ and hence

$$\mathbf{a}=a_x\mathbf{i}+a_y\mathbf{j}+a_z\mathbf{k}. \tag{2}$$

This equation shows the connection with the analysis of vector quantities by non-vectorial means, for a_x, a_y, a_z are the Cartesian co-ordinates of A with respect to the rectangular axes at O. We call a_x, a_y, a_z the *components of* **a**.

By Pythagoras' theorem $OA^2=OB^2+OC^2+OD^2$ and hence

$$a^2=a_x^2+a_y^2+a_z^2.$$

In applications of vector algebra to problems in geometry a useful concept is the *position vector* of a point, the position vector of A being $\overrightarrow{OA}=\mathbf{a}$, where O is some origin. Then $\overrightarrow{AB}=\overrightarrow{OB}-\overrightarrow{OA}=\mathbf{b}-\mathbf{a}$.

Example 1. *If* $\mathbf{a}=8\mathbf{i}-4\mathbf{j}+\mathbf{k}$, *find* a *and* $\hat{\mathbf{a}}$.

$a^2=64+16+1=81$, and so $a=9$. $\hat{\mathbf{a}}=\frac{1}{a}\mathbf{a}=\frac{8}{9}\mathbf{i}-\frac{4}{9}\mathbf{j}+\frac{1}{9}\mathbf{k}$.

Example 2. *Find the position vector of* P *which divides* AB *in the ratio* k : l.

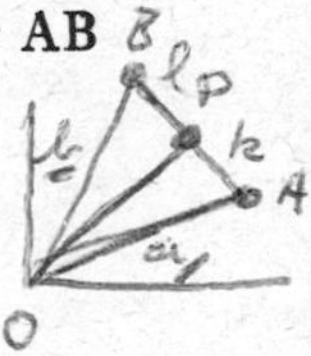

$$\overrightarrow{AP}=\frac{k}{k+l}\overrightarrow{AB}=\frac{k}{k+l}(\mathbf{b}-\mathbf{a}),$$

$$\overrightarrow{OP}=\overrightarrow{OA}+\overrightarrow{AP}=\mathbf{a}+\frac{k}{k+l}(\mathbf{b}-\mathbf{a})=\frac{l\mathbf{a}+k\mathbf{b}}{k+l}.$$

Example 3. *Find the equation of the straight line through* A *and* B.

This follows from Example 2, on replacing P by a variable point R on AB, so that the value of k/l takes an arbitrary value. The equation of the line is

$$\mathbf{r}=\mathbf{a}+\lambda(\mathbf{b}-\mathbf{a})=(1-\lambda)\mathbf{a}+\lambda\mathbf{b},$$

where λ is a variable parameter. Another form for the line is

$$\mathbf{r}=\mathbf{a}+\lambda\mathbf{t}$$

where $\mathbf{t}$ is a vector in the direction of the line. The components of this equation give the line in Cartesian co-ordinates as

$$x=a_x+\lambda t_x,\ y=a_y+\lambda t_y,\ z=a_z+\lambda t_z,$$

if $\mathbf{r}=x\mathbf{i}+y\mathbf{j}+z\mathbf{k}$. (If two vectors are equal their corresponding components must be equal, since Fig. 4 is the same in each case.)

We may equate the expressions for λ given by these equations, and write the equation of the line in the form

$$\frac{x-a_x}{t_x}=\frac{y-a_y}{t_y}=\frac{z-a_z}{t_z}.$$

Example 4. *Show that the medians of the triangle* ABC *are concurrent.*

By Example 2, the midpoint D of BC has position vector $\frac{1}{2}(\mathbf{b}+\mathbf{c})$. The point X on AD such that $AX=2XD$ has position vector $\frac{1}{3}\mathbf{a}+\frac{2}{3}\{\frac{1}{2}(\mathbf{b}+\mathbf{c})\}=\frac{1}{3}(\mathbf{a}+\mathbf{b}+\mathbf{c})$.

By symmetry, X also lies on the other two medians of the triangle.

1.3. THE SCALAR PRODUCT

So far all our results have come from adding vectors by means of the parallelogram law. But there are other ways in which combinations of vectors occur in physical applications. Consider the work done by a force $\mathbf{F}$ in a displacement given by the vector $\mathbf{s}$. As will be shown in section 2.3, the work done is the product of the magnitude of $\mathbf{F}$ and the component of $\mathbf{s}$ in the direction of $\mathbf{F}$. This combination of vectors is a product, since it involves the product of the moduli of $\mathbf{F}$ and $\mathbf{s}$, and it is a scalar quantity. Consequently we introduce the following definition.

The *scalar product* of two vectors $\mathbf{a}$ and $\mathbf{b}$, whose directions are inclined at an angle θ, is

$$\mathbf{a} \cdot \mathbf{b}=ab \cos \theta. \quad (3)$$

This scalar product is read as '*a* dot *b*'. To avoid possible confusion in complicated expressions, it is very desirable to be meticulous in inserting the dot whenever a scalar product is written.

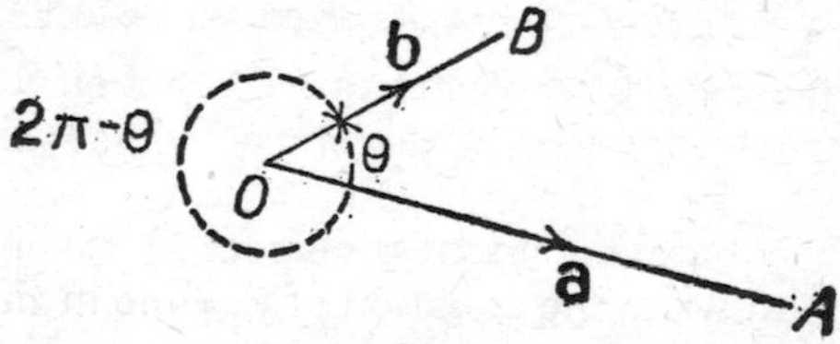

Figure 5

It is immaterial whether the angle between the vectors is taken to be the angle θ in Fig. 5 or the angle $2\pi-\theta$ measured in the opposite sense, for the cosine is unaltered, but it is essential that the angle shall be measured in a plane containing the directions of **a** and **b**. Since $\mathbf{a}\cdot\mathbf{b}=\mathbf{b}\cdot\mathbf{a}$ the commutative law is obeyed by scalar products.

From equation (3), the scalar product is the modulus of one vector multiplied by the projection on it of the other vector. The work done by the force **F** in a displacement **s** is therefore $\mathbf{F}\cdot\mathbf{s}$. Also the component of **a** in the direction of **b** is

$$a\cos\theta=\frac{1}{b}\mathbf{a}\cdot\mathbf{b}=\mathbf{a}\cdot\hat{\mathbf{b}}. \tag{4}$$

The equation $\mathbf{a}\cdot\mathbf{b}=0$ does not imply that either **a** or **b** must be zero. An alternative possibility is $\cos\theta=0$, which is true if **a** and **b** are perpendicular to each other. In fact $\mathbf{a}\cdot\mathbf{b}=0$ is most usefully thought of as the condition that two non-zero vectors **a** and **b** shall be perpendicular. If **a** and **b** are in the same direction $\mathbf{a}\cdot\mathbf{b}=ab$, and if they have opposite directions $\mathbf{a}\cdot\mathbf{b}=-ab$. In particular $\mathbf{a}\cdot\mathbf{a}$, which is written as $\mathbf{a}^2$, the square of the vector **a**, is given by

$$\mathbf{a}^2=\mathbf{a}\cdot\mathbf{a}=a^2. \tag{5}$$

When applied to the unit vectors in the co-ordinate directions, these results show that

$$\mathbf{i}^2=\mathbf{j}^2=\mathbf{k}^2=1,\ \mathbf{i}\cdot\mathbf{j}=\mathbf{j}\cdot\mathbf{k}=\mathbf{k}\cdot\mathbf{i}=0.$$

The distributive law for scalar products

$$\mathbf{a}\cdot(\mathbf{b}+\mathbf{c})=\mathbf{a}\cdot\mathbf{b}+\mathbf{a}\cdot\mathbf{c}$$

is true since the projection of $\mathbf{b}+\mathbf{c}$ on **a** is equal to the sum of the projections of **b** and **c**. Thus in Fig. 2, if we let D' and E' be the feet of the perpendiculars from D and E to the line OA, $AE'=AD'+D'E'$ and

$$\mathbf{a}\cdot(\mathbf{b}+\mathbf{c})=OA.AE'=OA.AD'+OA.D'E'=\mathbf{a}\cdot\mathbf{b}+\mathbf{a}\cdot\mathbf{c}.$$

No associative law is possible. It would say that $\mathbf{a}\cdot(\mathbf{b}\cdot\mathbf{c})$ was equal to $(\mathbf{a}\cdot\mathbf{b})\cdot\mathbf{c}$, yet neither of these expressions has a meaning. The bracketed terms are scalars, and so cannot be used in a product of vectors. An expression that has a meaning is $(\mathbf{a} \cdot \mathbf{b})\mathbf{c}$, this being a vector parallel to $\mathbf{c}$ of modulus $\pm(\mathbf{a} \cdot \mathbf{b})c$ (the sign being that required to make the modulus positive).

Multiplying terms in a product by a scalar $k>0$ does not affect the angle θ, and therefore

$$\mathbf{a} \cdot (k\mathbf{b}) = kab \cos\theta = k(\mathbf{a} \cdot \mathbf{b}) = (k\mathbf{a}) \cdot \mathbf{b}.$$

If $k<0$ the modulus of $k\mathbf{b}$ is $-kb$, but since $k\mathbf{b}$ is in the opposite direction to $\mathbf{b}$ the angle between $\mathbf{a}$ and $k\mathbf{b}$ is $\pi+\theta$. From (3) the above equation is hence true for all values of k, positive or negative.

These results show that the scalar product of $\mathbf{a}$ and $\mathbf{b}$ in terms of their components is given by

$$\begin{aligned}\mathbf{a} \cdot \mathbf{b} &= (a_x\mathbf{i}+a_y\mathbf{j}+a_z\mathbf{k}) \cdot (b_x\mathbf{i}+b_y\mathbf{j}+b_z\mathbf{k}) \\ &= a_xb_x+a_yb_y+a_zb_z. \end{aligned} \tag{6}$$

With $\mathbf{b}=\mathbf{a}$, (6) gives Pythagoras' theorem in three dimensions

$$a^2 = a_x^2+a_y^2+a_z^2, \tag{7}$$

expressing the modulus of $\mathbf{a}$ in terms of its components. This shows that $\mathbf{a}$ can only be zero if all its components are zero.

If the direction of $\mathbf{a}$ makes angles α, β, γ with the axes of x, y, z then from Fig. 4, $OB=OA \cos\alpha$ or $a_x=a\cos\alpha$, and similarly $a_y=a\cos\beta$, $a_z=a\cos\gamma$.

Equation (2) may be written as

$$\mathbf{a} = a(\cos\alpha\,\mathbf{i}+\cos\beta\,\mathbf{j}+\cos\gamma\,\mathbf{k}) \tag{8}$$

and hence by equation (1)

$$\hat{\mathbf{a}} = \cos\alpha\,\mathbf{i}+\cos\beta\,\mathbf{j}+\cos\gamma\,\mathbf{k},$$

which gives the direction of $\mathbf{a}$ in terms of $\cos\alpha$, $\cos\beta$, $\cos\gamma$, the *direction cosines* of OA. These directions are not independent of one another; from (7) and (8)

$$\cos^2\alpha+\cos^2\beta+\cos^2\gamma=1.$$

If $\mathbf{b}$ has direction cosines $\cos\lambda$, $\cos\mu$, $\cos\nu$ then from (3) and (6) the angle θ between $\mathbf{a}$ and $\mathbf{b}$ is given by

$$\cos\theta = \cos\alpha\cos\lambda+\cos\beta\cos\mu+\cos\gamma\cos\nu.$$

Example 1. *Find the angle between* $\mathbf{a}=2\mathbf{i}+\mathbf{j}+2\mathbf{k}$ *and* $\mathbf{b}=4\mathbf{i}-3\mathbf{j}$.

$$\mathbf{a}\cdot\mathbf{b}=8-3+0=5;\ a^2=4+1+4=9,\ a=3;$$
$$b^2=16+9=25,\ b=5.$$

Hence $\cos\theta=(\mathbf{a}\cdot\mathbf{b})/ab=\frac{1}{3}$, $\theta=70^\circ\ 32'$.

Example 2. *Show that the altitudes of a triangle* ABC *are concurrent.*

Let H be the intersection of the altitudes from A and B. Use position vectors. Since AH, BH are perpendicular to BC, CA,

$$(\mathbf{h}-\mathbf{a})\cdot(\mathbf{c}-\mathbf{b})=0,\quad \mathbf{h}\cdot(\mathbf{c}-\mathbf{b})=\mathbf{a}\cdot\mathbf{c}-\mathbf{a}\cdot\mathbf{b},$$
$$(\mathbf{h}-\mathbf{b})\cdot(\mathbf{a}-\mathbf{c})=0,\quad \mathbf{h}\cdot(\mathbf{a}-\mathbf{c})=\mathbf{b}\cdot\mathbf{a}-\mathbf{b}\cdot\mathbf{c}.$$

Adding these results, $\mathbf{h}\cdot(\mathbf{a}-\mathbf{b})=\mathbf{a}\cdot\mathbf{c}-\mathbf{b}\cdot\mathbf{c}=(\mathbf{a}-\mathbf{b})\cdot\mathbf{c}$ and hence $(\mathbf{h}-\mathbf{c})\cdot(\mathbf{a}-\mathbf{b})=0$. This shows that H is also on the altitude from C.

Example 3. *Find the equation of the sphere centre* A *and radius* c.

Let R be a point on the sphere. AR has length c. Hence

$$|\mathbf{r}-\mathbf{a}|=c,\ |\mathbf{r}-\mathbf{a}|^2=(\mathbf{r}-\mathbf{a})^2=r^2-2\mathbf{a}\cdot\mathbf{r}+a^2=c^2.$$

This is the required equation. In terms of components, it is

$$x^2+y^2+z^2-2a_xx-2a_yy-2a_zz+a_x{}^2+a_y{}^2+a_z{}^2-c^2=0.$$

Example 4. *Find the equation of a plane.*

Let P be the foot of the perpendicular from the origin O on to the plane, and let $\hat{\mathbf{n}}$ be the unit normal perpendicular to the plane. The point R is on the plane if $\overrightarrow{PR}$ is perpendicular to $\hat{\mathbf{n}}$, i.e. if $(\mathbf{r}-\mathbf{p})\cdot\hat{\mathbf{n}}=0$, $\mathbf{r}\cdot\hat{\mathbf{n}}=\mathbf{p}\cdot\hat{\mathbf{n}}$. Now $\mathbf{p}$ is parallel to $\hat{\mathbf{n}}$, and so the equation takes the form

$$\mathbf{r}\cdot\hat{\mathbf{n}}=p$$

where p is the distance from O to the plane, in the direction in which $\hat{\mathbf{n}}$ is measured. If $\hat{\mathbf{n}}$ has direction cosines $\cos\alpha$,

$\cos\beta$, $\cos\gamma$ this equation can be written, using (8), as

$$x\cos\alpha+y\cos\beta+z\cos\gamma=p.$$

1.4. THE VECTOR PRODUCT

There is another combination of two vectors which is of importance. The area of the parallelogram formed by the displacement vectors **a** and **b** in Fig. 1 or Fig. 5 is $S=ab\sin\theta$. This quantity is a product, since it depends on the product of the moduli of **a** and **b**, and we may also assign to it a direction, that of the normal to the plane of the parallelogram. We therefore make the following definition.

The *vector product* of two vectors **a** and **b** is

$$\mathbf{a}\wedge\mathbf{b}=ab\sin\theta\,\hat{\mathbf{n}} \tag{9}$$

where θ is the angle between the direction of **a** and the direction of **b**, and $\hat{\mathbf{n}}$ is a unit vector perpendicular to the plane of **a** and **b**. This vector product is read as '*a* cross *b*' or '*a* vec *b*'. Many authors do in fact denote a vector product by a cross sign, but when written this is liable to be confused with x, and the caret sign calls attention to the special properties of the vector product. Being defined as a scalar multiplied by a vector, $\mathbf{a}\wedge\mathbf{b}$ is a vector. Its modulus is the magnitude of one vector multiplied by the component perpendicular to it of the other.

There are still two possible directions for $\hat{\mathbf{n}}$. If in Fig. 5, **a** and **b** are considered to be in the plane of the paper, $\hat{\mathbf{n}}$ may be either up or down perpendicular to the paper. We must arbitrarily select one of these possibilities, and the convention universally adopted is that $\hat{\mathbf{n}}$ is given by the direction of advance of a right-handed screw turning from **a** to **b** in the sense in which θ is measured.

Thus in Fig. 5, $\hat{\mathbf{n}}$ is upwards. It would be possible tc measure from **a** to **b** in the opposite sense, with θ replaced by $2\pi-\theta$; then both $\sin\theta$ and $\hat{\mathbf{n}}$ would change sign, leaving equation (9) unaltered.

It follows from this definition that

$$\mathbf{a} \wedge \mathbf{b} = -\mathbf{b} \wedge \mathbf{a} \tag{10}$$

and thus the usual commutative law is not obeyed by vector products.

If **a** and **b** are parallel, possibly with opposite directions, then $\sin\theta=0$ and so $\mathbf{a} \wedge \mathbf{b}=0$. This then is the condition for two non-zero vectors to be parallel. In particular $\mathbf{a} \wedge \mathbf{a}=0$. For unit vectors in the co-ordinate directions forming a right-handed system (so that, for example, a right-handed screw turning from the x-direction to the y-direction advances in the z-direction) we have the following relations:

$$\mathbf{i} \wedge \mathbf{i} = \mathbf{j} \wedge \mathbf{j} = \mathbf{k} \wedge \mathbf{k} = 0,$$

$$\mathbf{j} \wedge \mathbf{k} = \mathbf{i} = -\mathbf{k} \wedge \mathbf{j},\ \mathbf{k} \wedge \mathbf{i} = \mathbf{j} = -\mathbf{i} \wedge \mathbf{k},\ \mathbf{i} \wedge \mathbf{j} = \mathbf{k} = -\mathbf{j} \wedge \mathbf{i}.$$

The distributive law for vector products

$$\mathbf{a} \wedge (\mathbf{b}+\mathbf{c}) = \mathbf{a} \wedge \mathbf{b} + \mathbf{a} \wedge \mathbf{c}$$

may be proved as follows. Let $\mathbf{b}'$, $\mathbf{c}'$ be the projections of **b**, **c** on to a plane perpendicular to **a**. Then we see from Fig. 6 that $\mathbf{b}'+\mathbf{c}'$ is the projection of $\mathbf{b}+\mathbf{c}$. Since only

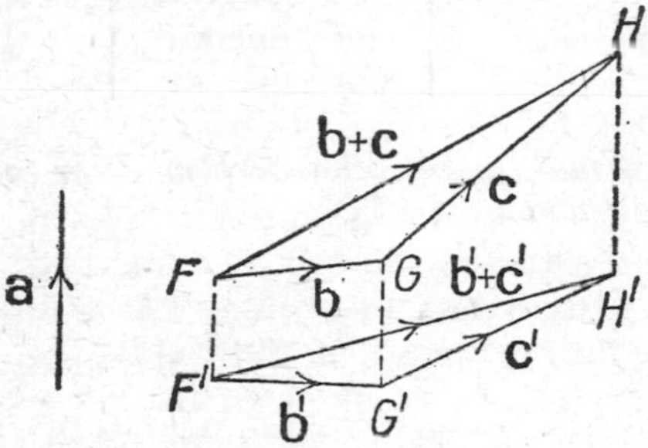

Figure 6

components perpendicular to **a** affect vector products with **a**,

$$\mathbf{a} \wedge \mathbf{b} = \mathbf{a} \wedge \mathbf{b}',\ \mathbf{a} \wedge \mathbf{c} = \mathbf{a} \wedge \mathbf{c}',\ \mathbf{a} \wedge (\mathbf{b}+\mathbf{c}) = \mathbf{a} \wedge (\mathbf{b}'+\mathbf{c}').$$

Now $\mathbf{a} \wedge \mathbf{b}'$ is perpendicular to $\mathbf{a}$ and $\mathbf{b}'$ and has modulus ab', i.e. it is obtained from $\mathbf{b}'$ by rotation through an angle $\frac{1}{2}\pi$ about the direction of $\mathbf{a}$, and magnification by the factor a. Applying this result also to $\mathbf{a} \wedge \mathbf{c}'$ and $\mathbf{a} \wedge (\mathbf{b}'+\mathbf{c}')$, we see that these three products form a triangle similar to $F'G'H'$, and so by the parallelogram law

$$\mathbf{a} \wedge (\mathbf{b}'+\mathbf{c}')=\mathbf{a} \wedge \mathbf{b}'+\mathbf{a} \wedge \mathbf{c}',$$

from which the distributive law follows.

The associative law, which would assert the equality of $\mathbf{a} \wedge (\mathbf{b} \wedge \mathbf{c})$ and $(\mathbf{a} \wedge \mathbf{b}) \wedge \mathbf{c}$, is false as we shall see in section 1.5, even though these expressions properly define vectors. Scalar multiplication is permissible. By the same arguments as for the scalar product,

$$\mathbf{a} \wedge (k\mathbf{b})=kab \sin \theta\, \hat{\mathbf{n}}=k(\mathbf{a} \wedge \mathbf{b})=(k\mathbf{a}) \wedge \mathbf{b}.$$

From these results the vector product of $\mathbf{a}$ and $\mathbf{b}$ in terms of their components is

$$\begin{aligned}\mathbf{a} \wedge \mathbf{b}&=(a_x\mathbf{i}+a_y\mathbf{j}+a_z\mathbf{k}) \wedge (b_x\mathbf{i}+b_y\mathbf{j}+b_z\mathbf{k})\\ &=(a_yb_z-a_zb_y)\mathbf{i}+(a_zb_x-a_xb_z)\mathbf{j}+(a_xb_y-a_yb_x)\mathbf{k}. \quad (11)\end{aligned}$$

This may be written in a symmetrical form as a determinant

$$\mathbf{a} \wedge \mathbf{b}=\begin{vmatrix} \mathbf{i} & \mathbf{j} & \mathbf{k} \\ a_x & a_y & a_z \\ b_x & b_y & b_z \end{vmatrix}.$$

Example 1. *Find a unit vector perpendicular to* $\mathbf{a}=2\mathbf{i}+3\mathbf{j}-4\mathbf{k}$ *and* $\mathbf{b}=\mathbf{i}+2\mathbf{j}$.

$$\mathbf{a} \wedge \mathbf{b}=(0+8)\mathbf{i}+(-4+0)\mathbf{j}+(4-3)\mathbf{k}=8\mathbf{i}-4\mathbf{j}+\mathbf{k},$$

$|\mathbf{a} \wedge \mathbf{b}|^2=64+16+1=81$, $|\mathbf{a} \wedge \mathbf{b}|=9$. The required vector is

$$(\mathbf{a} \wedge \mathbf{b})/|\mathbf{a} \wedge \mathbf{b}|=\tfrac{1}{9}(8\mathbf{i}-4\mathbf{j}+\mathbf{k}).$$

Example 2. *Find the area of the triangle* ABC.

The area is half that of the parallelogram given by AB and AC. Use position vectors.

$$\begin{aligned}\text{Area of triangle}&=\tfrac{1}{2}\overrightarrow{AB} \wedge \overrightarrow{AC}=\tfrac{1}{2}(\mathbf{b}-\mathbf{a}) \wedge (\mathbf{c}-\mathbf{a})\\ &=\tfrac{1}{2}(\mathbf{b} \wedge \mathbf{c}-\mathbf{a} \wedge \mathbf{c}-\mathbf{b} \wedge \mathbf{a}+\mathbf{a} \wedge \mathbf{a})\\ &=\tfrac{1}{2}(\mathbf{b} \wedge \mathbf{c}+\mathbf{c} \wedge \mathbf{a}+\mathbf{a} \wedge \mathbf{b}).\end{aligned}$$

Example 3. *Show that the sum of the vector areas, taken outwards, of the faces of the tetrahedron OABC is zero.*

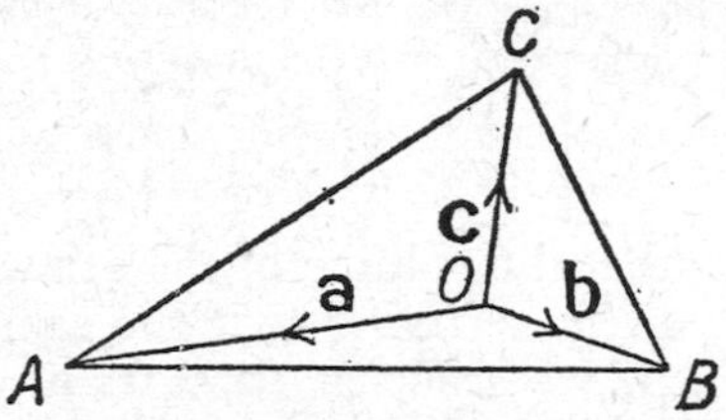

Figure 7

The area of the face *ABC* is as found in Example 2, if *O* is considered to be below the plane of *ABC* in Fig. 7. The three terms of the final expression are the areas of the faces *OBC*, *OCA*, *OAB* measured inwards. Hence the result, which confirms that vector areas may be added by the vector law, satisfying Condition 2.

1.5. TRIPLE PRODUCTS

Since $\mathbf{a} \cdot \mathbf{b}$ is a scalar, no further vector operations can be applied to it. But $\mathbf{a} \wedge \mathbf{b}$ is a vector, and we may consider its scalar or vector product with another vector $\mathbf{c}$.

The *triple scalar product* $(\mathbf{a} \wedge \mathbf{b}) \cdot \mathbf{c}$ is a scalar quantity, being the scalar product of $\mathbf{a} \wedge \mathbf{b}$ and $\mathbf{c}$. To demonstrate its significance, consider the parallelepiped shown in Fig. 8. The modulus of $\mathbf{a} \wedge \mathbf{b}$ is the area S of the face $OAFB$, and its direction $\hat{\mathbf{n}}$ is normal to the face. Then

$$(\mathbf{a} \wedge \mathbf{b}) \cdot \mathbf{c} = cS \cos \theta = V,$$

where θ is the angle between $\mathbf{c}$ and $\hat{\mathbf{n}}$. Since $c \cos \theta$ is the perpendicular height, V is the volume of the parallelepiped, as may be shown by the same type of argument as is used to derive the area of a parallelogram. In terms of components, from (11),

$$(\mathbf{a} \wedge \mathbf{b}) \cdot \mathbf{c}$$

$$=a_y b_z c_x - a_z b_y c_x + a_z b_x c_y - a_x b_z c_y + a_x b_y c_z - a_y b_x c_z \quad (12)$$

$$= \begin{vmatrix} a_x & a_y & a_z \\ b_x & b_y & b_z \\ c_x & c_y & c_z \end{vmatrix}.$$

By observing that the volume of the same parallelepiped is represented, or by verifying that the same six terms in (12) are obtained, we see that $(\mathbf{b} \wedge \mathbf{c}) \cdot \mathbf{a}$ and $(\mathbf{c} \wedge \mathbf{a}) \cdot \mathbf{b}$ are equal to $(\mathbf{a} \wedge \mathbf{b}) \cdot \mathbf{c}$. The vectors in a scalar product

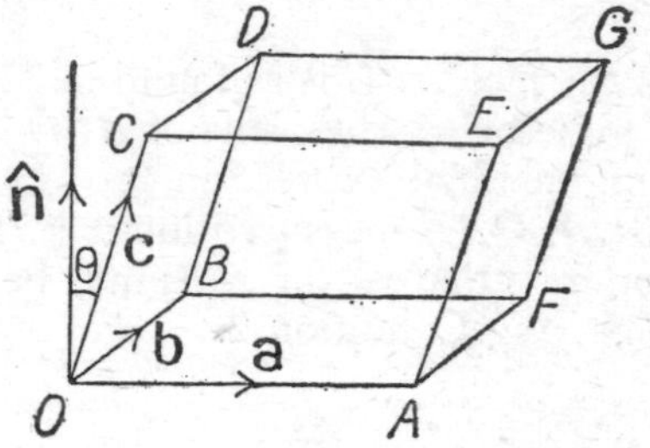

Figure 8

may be interchanged, hence $(\mathbf{a} \wedge \mathbf{b}) \cdot \mathbf{c}=\mathbf{c} \cdot (\mathbf{a} \wedge \mathbf{b})$, but an interchange in a vector product causes a change of sign, and hence $(\mathbf{a} \wedge \mathbf{b}) \cdot \mathbf{c}= -(\mathbf{b} \wedge \mathbf{a}) \cdot \mathbf{c}$. Combining all these results, we can relate the twelve possible arrangements of the triple product as follows:

$$\begin{aligned} V&=\mathbf{a} \cdot (\mathbf{b} \wedge \mathbf{c})=(\mathbf{b} \wedge \mathbf{c}) \cdot \mathbf{a}= -\mathbf{a} \cdot (\mathbf{c} \wedge \mathbf{b})= -(\mathbf{c} \wedge \mathbf{b}) \cdot \mathbf{a} \\ &=\mathbf{b} \cdot (\mathbf{c} \wedge \mathbf{a})=(\mathbf{c} \wedge \mathbf{a}) \cdot \mathbf{b}= -\mathbf{b} \cdot (\mathbf{a} \wedge \mathbf{c})= -(\mathbf{a} \wedge \mathbf{c}) \cdot \mathbf{b} \\ &=\mathbf{c} \cdot (\mathbf{a} \wedge \mathbf{b})=(\mathbf{a} \wedge \mathbf{b}) \cdot \mathbf{c}= -\mathbf{c} \cdot (\mathbf{b} \wedge \mathbf{a})= -(\mathbf{b} \wedge \mathbf{a}) \cdot \mathbf{c}. \end{aligned} \quad (13)$$

We see that the value does not depend on the positions of the dot and caret signs. In the positive terms **a**, **b**, **c** occur in cyclic order, while in the negative terms they are in non-cyclic order. The triple scalar product may conveniently be written as [**a**, **b**, **c**]. This is sufficient, as it gives the three vectors and their cyclic order. Thus, for example,

$$[\mathbf{a}, \mathbf{b}, \mathbf{c}]=[\mathbf{b}, \mathbf{c}, \mathbf{a}]=-[\mathbf{a}, \mathbf{c}, \mathbf{b}].$$

An important property of the triple scalar product is that it is zero if the vectors are coplanar, for then the volume $V=0$. In particular it is zero if two of the vectors are equal, as is seen immediately on writing $[\mathbf{a}, \mathbf{a}, \mathbf{c}]=(\mathbf{a} \wedge \mathbf{a}) \cdot \mathbf{c}=0$.

The *triple vector product* $(\mathbf{a} \wedge \mathbf{b}) \wedge \mathbf{c}=\mathbf{v}$ is a vector perpendicular to both **c** and $\mathbf{a} \wedge \mathbf{b}$. It hence lies in the plane of **a** and **b**, and we can write $\mathbf{v}=l\mathbf{a}+m\mathbf{b}$. The values of l and m may be found by using (11) twice to obtain

$$\begin{aligned}\mathbf{v}=&\{(a_z b_x-a_x b_z)c_z-(a_x b_y-a_y b_x)c_y\}\mathbf{i}\\ &+\{(a_x b_y-a_y b_x)c_x-(a_y b_z-a_z b_y)c_z\}\mathbf{j}\\ &+\{(a_y b_z-a_z b_y)c_y-(a_z b_x-a_x b_z)c_x\}\mathbf{k}\\ =&-(a_x\mathbf{i}+a_y\mathbf{j}+a_z\mathbf{k})(b_x c_x+b_y c_y+b_z c_z)\\ &+(b_x\mathbf{i}+b_y\mathbf{j}+b_z\mathbf{k})(a_x c_x+a_y c_y+a_z c_z).\end{aligned}$$

Hence $l=-\mathbf{b} \cdot \mathbf{c}$, $m=\mathbf{a} \cdot \mathbf{c}$, and we have

$$(\mathbf{a} \wedge \mathbf{b}) \wedge \mathbf{c}=(\mathbf{a} \cdot \mathbf{c})\mathbf{b}-(\mathbf{b} \cdot \mathbf{c})\mathbf{a}. \tag{14}$$

Now $(\mathbf{a} \wedge \mathbf{b}) \wedge \mathbf{c}=-\mathbf{c} \wedge (\mathbf{a} \wedge \mathbf{b})$ by (10), and on permuting **a**, **b**, **c** we obtain from equation (14)

$$\mathbf{a} \wedge (\mathbf{b} \wedge \mathbf{c})=(\mathbf{a} \cdot \mathbf{c})\mathbf{b}-(\mathbf{a} \cdot \mathbf{b})\mathbf{c}. \tag{15}$$

From (14) and (15) we see at once that the bracket in the triple vector product cannot be omitted without causing ambiguity. As an aid to remembering these expressions, note that there is no component in the direction of the unbracketed vector, and that the first scalar product, with a positive sign, is formed from the outer pair of vectors.

It is sometimes useful to express a vector **a** as the sum of two component vectors, one parallel and one perpendicular to another vector **b**. From (4) the component parallel to **b** is $(\mathbf{a} \cdot \hat{\mathbf{b}})\hat{\mathbf{b}}=(1/b^2)(\mathbf{a} \cdot \mathbf{b})\mathbf{b}$, and so the perpendicular component is

$$\mathbf{a}-\frac{1}{b^2}(\mathbf{a}\cdot\mathbf{b})\mathbf{b}=\frac{1}{b^2}\{b^2\mathbf{a}-(\mathbf{a}\cdot\mathbf{b})\mathbf{b}\}=\frac{1}{b^2}\mathbf{b}\wedge(\mathbf{a}\wedge\mathbf{b}).$$

Hence

$$\mathbf{a}=\frac{1}{b^2}\{(\mathbf{a}\cdot\mathbf{b})\mathbf{b}+\mathbf{b}\wedge(\mathbf{a}\wedge\mathbf{b})\}. \tag{16}$$

Products of more than three vectors can readily be written down, and can be manipulated by use of the formulae already established, but the results are hardly worth remembering.

Example 1. $[\mathbf{a}, \mathbf{b}, \mathbf{c}]\mathbf{d}=[\mathbf{d}, \mathbf{b}, \mathbf{c}]\mathbf{a}+[\mathbf{a}, \mathbf{d}, \mathbf{c}]\mathbf{b}+[\mathbf{a}, \mathbf{b}, \mathbf{d}]\mathbf{c}$.

Considering first $(\mathbf{a}\wedge\mathbf{b})$ and then $(\mathbf{c}\wedge\mathbf{d})$ as a single vector, we have

$$\begin{aligned}(\mathbf{a}\wedge\mathbf{b})\wedge(\mathbf{c}\wedge\mathbf{d})&=[\mathbf{a}, \mathbf{b}, \mathbf{d}]\mathbf{c}-[\mathbf{a}, \mathbf{b}, \mathbf{c}]\mathbf{d}\\&=[\mathbf{a}, \mathbf{c}, \mathbf{d}]\mathbf{b}-[\mathbf{b}, \mathbf{c}, \mathbf{d}]\mathbf{a},\end{aligned}$$

from which the result follows. This useful formula expresses **d** as the sum of three vectors in the directions of **a**, **b**, **c** and shows that this is always possible provided $[\mathbf{a}, \mathbf{b}, \mathbf{c}]\neq 0$, i.e. provided **a**, **b**, **c** are non-zero and not coplanar.

Example 2. *Find the equation of the plane through* O, A, B.

Use position vectors from O. If R is a point in the plane, **r**, **a**, **b** are coplanar and hence $[\mathbf{r}, \mathbf{a}, \mathbf{b}]=0$. This is the required equation since it may be written as $\mathbf{r}\cdot(\mathbf{a}\wedge\mathbf{b})=0$, and this represents a plane as shown in Example 4 of section 1.3. An alternative form of equation for the plane is $\mathbf{r}=\lambda\mathbf{a}+\mu\mathbf{b}$, where λ and μ are variable parameters, since **r** may be resolved into components parallel to **a** and **b**.

Example 3. *Find the condition for the lines* $\mathbf{r}=\mathbf{a}+\lambda\mathbf{b}$ *and* $\mathbf{r}=\mathbf{c}+\mu\mathbf{d}$ *to intersect.*

At the point of intersection $\mathbf{c}-\mathbf{a}=\lambda\mathbf{b}-\mu\mathbf{d}$. Hence $\mathbf{c}-\mathbf{a}$ lies in the plane containing **b** and **d**, and the required condition is $[\mathbf{c}-\mathbf{a}, \mathbf{b}, \mathbf{d}]=0$.

Example 4. *Solve for* $\mathbf{x}$ *the equation* $\mathbf{x} \wedge \mathbf{a} = \mathbf{b}$.

The equation requires $\mathbf{b}$ to be perpendicular to $\mathbf{a}$; if this is not so no solution exists. Now $\mathbf{a}$, $\mathbf{b}$, $\mathbf{a} \wedge \mathbf{b}$ are mutually perpendicular. Expressing $\mathbf{x}$ in terms of components in these directions, and noting that $\mathbf{x}$ is perpendicular to $\mathbf{b}$, we write $\mathbf{x} = \lambda\mathbf{a} + \mu(\mathbf{a} \wedge \mathbf{b})$. Substituting into the equation we find $\mathbf{b} = \mathbf{x} \wedge \mathbf{a} = \mu(\mathbf{a} \wedge \mathbf{b}) \wedge \mathbf{a} = \mu a^2\mathbf{b}$, since $\mathbf{a} \cdot \mathbf{b} = 0$, and so $\mu = 1/a^2$ and $\mathbf{x} = \lambda\mathbf{a} + (1/a^2)\mathbf{a} \wedge \mathbf{b}$.

Now $\{\lambda\mathbf{a} + (1/a^2)\mathbf{a} \wedge \mathbf{b}\} \wedge \mathbf{a} = \mathbf{b}$ for any value of λ, and hence this expression for $\mathbf{x}$ is the general solution of the equation.

1.6. DIFFERENTIATION AND INTEGRATION

In many physical applications the vector quantities in which we are interested will not be constant but will vary with time. We are thus led to consider the properties of a vector $\mathbf{a}$ which is a continuously varying function of a scalar variable t. The differential coefficient of $\mathbf{a}(t)$ with respect to t may be defined by precisely the same procedure as that used to define the differential coefficient of a scalar quantity. Consider the ratio

$$\frac{\mathbf{a}(t+\delta t) - \mathbf{a}(t)}{\delta t} = \frac{\delta \mathbf{a}}{\delta t}.$$

The ratio is a vector, if δt is any non-zero change in t. If $\delta\mathbf{a}/\delta t$ tends to a limit as $\delta t \to 0$, we define this limit as $d\mathbf{a}/dt$ or $\dot{\mathbf{a}}$, the differential coefficient of $\mathbf{a}(t)$ with respect to t. It is clearly a vector.

Consider the motion of a point P which travels along a curve and is at A at time t and at A' at time $t+\delta t$, as shown in Fig. 9. Then $\overrightarrow{OA} = \mathbf{a}(t)$, $\overrightarrow{OA'} = \mathbf{a}(t+\delta t)$ and hence $\overrightarrow{AA'} = \delta\mathbf{a}$. The magnitude of $\delta\mathbf{a}/\delta t$ is $AA'/\delta t$. The limit $\dot{\mathbf{a}}$ is called the *velocity* of P. Its modulus is the *speed* (the rate of change of position of P, a scalar quantity) and its direction is that of the tangent to the curve on which P moves. In fact if a point has position vector $\mathbf{r}$ we have shown that its velocity is a vector $\mathbf{v}=\dot{\mathbf{r}}$. Repeating the argument, we see that the rate of change of the velocity is a vector, the *acceleration* $\mathbf{f}=\dot{\mathbf{v}}=\ddot{\mathbf{r}}$.

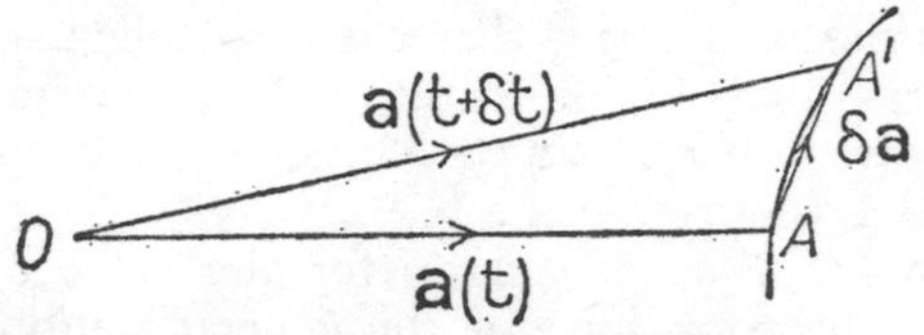

Figure 9

Formulae for the differentiation of sums and products of vectors are deduced in just the same way as for scalar functions, and have the expected forms. Care is needed to preserve the order where vector products are concerned. We may note the following results:

$$\frac{d}{dt}(\mathbf{a}+\mathbf{b})=\frac{d\mathbf{a}}{dt}+\frac{d\mathbf{b}}{dt},$$

$$\frac{d}{dt}(k\mathbf{a})=\frac{dk}{dt}\mathbf{a}+k\frac{d\mathbf{a}}{dt},$$

$$\frac{d}{dt}(\mathbf{a}\cdot\mathbf{b})=\frac{d\mathbf{a}}{dt}\cdot\mathbf{b}+\mathbf{a}\cdot\frac{d\mathbf{b}}{dt},$$

$$\frac{d}{dt}(\mathbf{a}\wedge\mathbf{b})=\frac{d\mathbf{a}}{dt}\wedge\mathbf{b}+\mathbf{a}\wedge\frac{d\mathbf{b}}{dt},$$

$$\frac{d}{dt}[\mathbf{a},\mathbf{b},\mathbf{c}]=\left[\frac{d\mathbf{a}}{dt},\mathbf{b},\mathbf{c}\right]+\left[\mathbf{a},\frac{d\mathbf{b}}{dt},\mathbf{c}\right]+\left[\mathbf{a},\mathbf{b},\frac{d\mathbf{c}}{dt}\right].$$

As an illustration, consider $\mathbf{a}\wedge\mathbf{b}$. Its value at $t+\delta t$ is $(\mathbf{a}+\delta\mathbf{a})\wedge(\mathbf{b}+\delta\mathbf{b})=\mathbf{a}\wedge\mathbf{b}+\delta\mathbf{a}\wedge\mathbf{b}+\mathbf{a}\wedge\delta\mathbf{b}+\delta\mathbf{a}\wedge\delta\mathbf{b}$ and hence

$$\frac{(\mathbf{a}\wedge\mathbf{b})(t+\delta t)-(\mathbf{a}\wedge\mathbf{b})(t)}{\delta t}=\frac{\delta\mathbf{a}}{\delta t}\wedge\mathbf{b}+\mathbf{a}\wedge\frac{\delta\mathbf{b}}{\delta t}+\frac{\delta\mathbf{a}}{\delta t}\wedge\delta\mathbf{b}.$$

Taking the limit as $\delta t\to 0$, we obtain the required formula.

In terms of components, $\mathbf{a}(t)=a_x(t)\mathbf{i}+a_y(t)\mathbf{j}+a_z(t)\mathbf{k}$ and hence

$$\frac{d\mathbf{a}}{dt}=\frac{da_x}{dt}\mathbf{i}+\frac{da_y}{dt}\mathbf{j}+\frac{da_z}{dt}\mathbf{k} \tag{17}$$

since $\mathbf{i}, \mathbf{j}, \mathbf{k}$ are constant vectors.

A result of particular interest is

$$\frac{d}{dt}\mathbf{a}^2=\frac{d}{dt}(\mathbf{a}\cdot\mathbf{a})=2\mathbf{a}\cdot\frac{d\mathbf{a}}{dt}.$$

Since $\mathbf{a}^2=a^2$ we see that

$$\mathbf{a}\cdot\frac{d\mathbf{a}}{dt}=a\frac{da}{dt}. \tag{18}$$

If $\mathbf{a}$ has constant modulus, $da/dt=0$. We cannot deduce that $d\mathbf{a}/dt$ is zero, only that it is perpendicular to $\mathbf{a}$. Thus in Fig. 9, P could move over a sphere with centre O, its speed being given by $|d\mathbf{a}/dt|$. It is important to note that $|d\mathbf{a}/dt|$ is not in general equal to da/dt, as this example shows.

The process of integrating a vector with respect to a scalar is equally straightforward. If $d\mathbf{a}/dt=\mathbf{b}(t)$, we may write

$$\mathbf{a}(t)=\int_{t_0}^{t}\mathbf{b}(\tau)d\tau+\mathbf{c},$$

where $\mathbf{c}$ is a constant vector of integration giving the value of $\mathbf{a}$ when $t=t_0$. In terms of components,

$$a_x(t)=\int_{t_0}^{t}b_x(\tau)d\tau+c_x,\quad a_y(t)=\int_{t_0}^{t}b_y(\tau)d\tau+c_y,$$

$$a_z(t)=\int_{t_0}^{t}b_z(\tau)d\tau+c_z,$$

which are just the usual formulae for the integrals of scalar functions. In particular if $\mathbf{v}$ is the velocity of a point, its position vector $\mathbf{r}$ is given by $\mathbf{r}(t)=\int_{t_0}^{t}\mathbf{v}(\tau)d\tau+\mathbf{r}_0$, where $\mathbf{r}_0$ is the position vector at time $t=t_0$.

The following example illustrates a variety of the vector

properties we have been studying. The result can also be proved in a straightforward manner by expressing the vectors in terms of their components. The details are left to the reader.

Example 1. *The acceleration* $\ddot{\mathbf{r}}$ *of a charged particle moving in a uniform magnetic field is given by* $\ddot{\mathbf{r}}=\mathbf{a}\wedge\dot{\mathbf{r}}$, *where* $\mathbf{a}$ *is constant. If at time* $t=0$ *the velocity is perpendicular to* $\mathbf{a}$, *prove that the particle moves in a circle with constant speed.*

The scalar product with $\dot{\mathbf{r}}$ gives $\ddot{\mathbf{r}}\cdot\dot{\mathbf{r}}=[\mathbf{a},\dot{\mathbf{r}},\dot{\mathbf{r}}]=0$. Integrating with respect to the time, $\dot{\mathbf{r}}^2=$constant, and hence the speed $|\dot{\mathbf{r}}|$ is constant.

Integrating the original equation, $\dot{\mathbf{r}}=\mathbf{a}\wedge\mathbf{r}+\mathbf{b}$, where $\mathbf{b}$ is constant. At $t=0$, $\dot{\mathbf{r}}$ and $\mathbf{a}\wedge\mathbf{r}$ are perpendicular to $\mathbf{a}$, hence $\mathbf{b}$ is perpendicular to $\mathbf{a}$ and we can write $\mathbf{b}=\mathbf{a}\wedge\mathbf{c}$, for some $\mathbf{c}$. Thus $\dot{\mathbf{r}}=\mathbf{a}\wedge(\mathbf{r}+\mathbf{c})$.

The scalar product of this equation with $\mathbf{r}+\mathbf{c}$ gives $(\mathbf{r}+\mathbf{c})\cdot\dot{\mathbf{r}}=0$, and therefore $(\mathbf{r}+\mathbf{c})^2=$constant; the scalar product with $\mathbf{a}$ gives $\mathbf{a}\cdot\dot{\mathbf{r}}=0$, and therefore $\mathbf{a}\cdot\mathbf{r}=$constant. These results show that the particle moves on a sphere and on a plane. Hence it must move on their intersection, which is a circle.

1.7. RELATIVE VELOCITY, ANGULAR VELOCITY

All the vector algebra which we require has now been developed, and we may turn to the consideration of mechanics. Before embarking on our systematic study of dynamics in the next chapter, we first shall consider a few direct applications of vector algebra.

The *relative displacement* $\mathbf{b}-\mathbf{a}$ of B relative to A is a vector, where $\overrightarrow{OA}=\mathbf{a}$, $\overrightarrow{OB}=\mathbf{b}$, and O is some origin. From section 1.6, the *relative velocity* $\dot{\mathbf{b}}-\dot{\mathbf{a}}$ of B relative to A is also a vector. Relative displacement and relative velocity are independent of the position and movement of the origin O, which as we shall see is of fundamental importance, for the idea of a 'fixed' origin is hardly a realistic one. Relative velocity is of practical use in many problems, of which the following is a simple example.

Example 1. *To a cyclist travelling due north along a straight road at* 10 *m.p.h., the wind appears to come from the east. When he increases his speed to* 15 *m.p.h. the wind appears to come from the direction N.*67°*E. Find the strength and direction of the wind over the ground.*

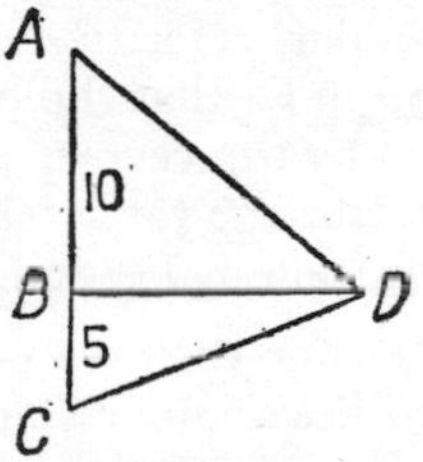

Figure 10

In Fig. 10, $\overrightarrow{DA}$ gives the true wind velocity, $\overrightarrow{BA}$, $\overrightarrow{CA}$ the cyclist's velocities, and $\overrightarrow{DB}$, $\overrightarrow{DC}$ the wind relative to the cyclist. $BD=5 \tan 67° = 11{\cdot}78$, the angle $BAD = \tan^{-1} 1{\cdot}178 = 49° \ 40'$, $AD = 10 \sec 49° \ 40' = 15{\cdot}81$. Hence the wind speed is 15·81 m.p.h. from the direction S.49° 40′E.

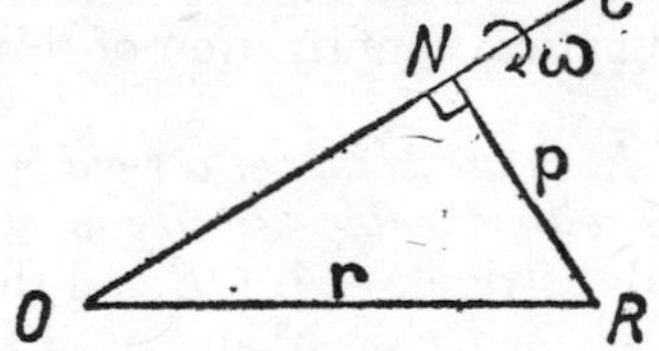

Figure 11

Consider a rigid body pivoted at O and rotating about the axis OC, a line fixed in space. We can specify the motion by the vector $\boldsymbol{\omega}$, with direction that of OC and magnitude the angular speed of the body in radians per second, in the sense of a right-handed screw; we say that the body rotates with *angular velocity* $\boldsymbol{\omega}$. The velocity of the point R of the body has magnitude ωp, where p is the length of the perpendicular RN from R to OC, and is in a direction perpendicular to the plane ROC; i.e. the velocity is

$$\mathbf{v}=\boldsymbol{\omega}\wedge\mathbf{r}. \tag{19}$$

Before we can say that angular velocity is a vector we must prove that angular velocities can be added by the parallelogram law. Suppose that the body has an additional angular velocity given by the vector $\boldsymbol{\omega}'$, about the axis OC'. Since velocity is a vector the total velocity of R is

$$\boldsymbol{\omega}\wedge\mathbf{r}+\boldsymbol{\omega}'\wedge\mathbf{r}=(\boldsymbol{\omega}+\boldsymbol{\omega}')\wedge\mathbf{r} \tag{20}$$

by the distributive law for vector products. The combined motion is that due to the angular velocity given by $\boldsymbol{\omega}+\boldsymbol{\omega}'$, the vector sum of $\boldsymbol{\omega}$ and $\boldsymbol{\omega}'$, and hence angular velocity is indeed a vector, satisfying Condition 2.

As remarked in section 1.1, *angular displacement* is not a vector, though it may be represented by a vector quantity $\mathbf{a}$ giving the angle turned and the axis of rotation. The displacement of the point R is not $\mathbf{a}\wedge\mathbf{r}$, and it cannot be proved, as from equation (20), that the effect of two angular displacements $\mathbf{a}$ and $\mathbf{a}'$ is given by their vector sum.

It may be thought to be intuitively obvious that the most general motion of a rigid body pivoted at O is an angular velocity about some axis through O. The following example provides mathematical confirmation of this.

Example 2. *If* A, B *are points of a rigid body pivoted at* O, *the body rotates with angular velocity* $\boldsymbol{\omega}=(\dot{\mathbf{a}}\wedge\dot{\mathbf{b}})/(\dot{\mathbf{a}}\cdot\mathbf{b})$.

Since the body is rigid, OA, OB and the angle between them are constant, hence $\mathbf{a}^2$, $\mathbf{b}^2$ and $\mathbf{a}\cdot\mathbf{b}$ are constant, and

$$\mathbf{a}\cdot\dot{\mathbf{a}}=0,\quad \mathbf{b}\cdot\dot{\mathbf{b}}=0,\quad \dot{\mathbf{a}}\cdot\mathbf{b}+\mathbf{a}\cdot\dot{\mathbf{b}}=0.$$

If the motion of A and B is due to an angular velocity $\boldsymbol{\omega}$, $\dot{\mathbf{a}}$ and $\dot{\mathbf{b}}$ must be perpendicular to $\boldsymbol{\omega}$, and hence $\boldsymbol{\omega}=\lambda(\dot{\mathbf{a}}\wedge\dot{\mathbf{b}})$, giving as the velocity of B

$$\boldsymbol{\omega}\wedge\mathbf{b}=\lambda(\dot{\mathbf{a}}\wedge\dot{\mathbf{b}})\wedge\mathbf{b}=\lambda(\dot{\mathbf{a}}\cdot\mathbf{b})\dot{\mathbf{b}}.$$

The velocity of B is given correctly as $\dot{\mathbf{b}}$ if $\lambda=1/(\dot{\mathbf{a}}\cdot\mathbf{b})$. Then

$$\boldsymbol{\omega}=\frac{\dot{\mathbf{a}}\wedge\dot{\mathbf{b}}}{\dot{\mathbf{a}}\cdot\mathbf{b}}=\frac{\dot{\mathbf{b}}\wedge\dot{\mathbf{a}}}{\dot{\mathbf{b}}\cdot\mathbf{a}},$$

from the relation proved above. By symmetry, the velocity of A due to $\boldsymbol{\omega}$ is given correctly as $\dot{\mathbf{a}}$. A rigid body's motion is completely determined by the motion of three of its points, and so the actual motion of the body is indeed just this angular velocity.

1.8. FORCES AND MOMENTS

That forces are vectors is a consequence of Newton's laws of motion, as will be discussed in section 2.2. Hence forces may be added by the vector law, which in this application is known as the *parallelogram of forces*. If we have a set of forces $\mathbf{F}_1, \mathbf{F}_2, \ldots, \mathbf{F}_n$ their sum

$$\mathbf{F}=\mathbf{F}_1+\mathbf{F}_2+\ldots+\mathbf{F}_n \tag{21}$$

is called the *resultant force*. For forces applied to a dynamical system, the rate of change of momentum is equal to $\mathbf{F}$, as shown in section 3.1. If the system is to be in equilibrium, it is necessary that $\mathbf{F}=0$. In this case the displacement vectors representing $\mathbf{F}_1, \mathbf{F}_2, \ldots, \mathbf{F}_n$ must form a closed polygon. When there are just three forces involved, the polygon reduces to the well-known triangle of forces.

The condition $\mathbf{F}=0$ is not in itself sufficient to ensure that equilibrium is possible. As shown in section 3.2, the rate of change of angular momentum, about a suitable point O, of the system to which the forces are applied is

$$\mathbf{G}=\mathbf{r}_1\wedge\mathbf{F}_1+\mathbf{r}_2\wedge\mathbf{F}_2+\ldots+\mathbf{r}_n\wedge\mathbf{F}_n, \tag{22}$$

where $\mathbf{r}_1, \mathbf{r}_2, \ldots, \mathbf{r}_n$ are the position vectors from O of the points of applications of the forces $\mathbf{F}_1, \mathbf{F}_2, \ldots, \mathbf{F}_n$. We call $\mathbf{r}_1 \wedge \mathbf{F}_1$ the *moment* of the force $\mathbf{F}_1$ about O. Equation (22) shows that moment of a force is a vector since $\mathbf{G}$, the *resultant moment*, is the vector sum of the individual moments. For equilibrium, $\mathbf{G}=0$ is another necessary condition. It will be shown in section 3.2 that for a rigid body $\mathbf{F}=0$ and $\mathbf{G}=0$ are together sufficient to ensure that equilibrium is possible. For forces applied at a single point $\mathbf{r}$,

$$\mathbf{G}=\mathbf{r} \wedge (\mathbf{F}_1+\mathbf{F}_2+\ldots+\mathbf{F}_n)=\mathbf{r} \wedge \mathbf{F}.$$

In this case $\mathbf{F}=0$ implies that $\mathbf{G}=0$. For extended systems a study of $\mathbf{G}$ is essential.

In terms of components, suppose that $\mathbf{r}_1=x_1\mathbf{i}+y_1\mathbf{j}+z_1\mathbf{k}$, $\mathbf{F}_1=X_1\mathbf{i}+Y_1\mathbf{j}+Z_1\mathbf{k}$. Then

$$\mathbf{G}_1=(y_1Z_1-z_1Y_1)\mathbf{i}+(z_1X_1-x_1Z_1)\mathbf{j}+(x_1Y_1-y_1X_1)\mathbf{k}. \quad (23)$$

For a two-dimensional force (X_1, Y_1) acting at (x_1, y_1) the z-component of (23) gives the counter-clockwise moment about O, as defined in elementary treatments of two-dimensional statics. But this is more satisfactorily interpreted as the moment of the force about a *line*, the z-axis, taken as usual in the right-handed screw sense. Neither z_1 nor Z_1 affect the moment about the z-axis. Furthermore, there is no special significance in the choice of the direction of the axes, so we can say that the component of $\mathbf{G}_1$ in any direction is equal to the moment of the force $\mathbf{F}_1$ about a line through O in that direction. The moment about a line in the direction $\hat{\mathbf{n}}$ through O is accordingly defined as

$$\hat{\mathbf{n}} \cdot (\mathbf{r}_1 \wedge \mathbf{F}_1)=[\hat{\mathbf{n}}, \mathbf{r}_1, \mathbf{F}_1].$$

The line through the point $\mathbf{r}_1$ parallel to $\mathbf{F}_1$ is given by $\mathbf{r}=\mathbf{r}_1+\lambda\mathbf{F}_1$, where λ is a variable parameter. Now

$$\mathbf{r} \wedge \mathbf{F}_1=\mathbf{r}_1 \wedge \mathbf{F}_1=\mathbf{G}_1$$

and thus the moment is unaltered if $\mathbf{F}_1$ acts at any point on this line, which is called the *line of action* of the force $\mathbf{F}_1$.

Consider the moments of the forces $\mathbf{F}_1, \mathbf{F}_2, \ldots, \mathbf{F}_n$

about a new centre O', where $\overrightarrow{OO'}=\mathbf{a}$. The sum of the moments is

$$\begin{aligned}\mathbf{G}'&=(\mathbf{r}_1-\mathbf{a})\wedge\mathbf{F}_1+(\mathbf{r}_2-\mathbf{a})\wedge\mathbf{F}_2+\ldots+(\mathbf{r}_n-\mathbf{a})\wedge\mathbf{F}_n\\&=\mathbf{r}_1\wedge\mathbf{F}_1+\mathbf{r}_2\wedge\mathbf{F}_2+\ldots+\mathbf{r}_n\wedge\mathbf{F}_n\\&\qquad-\mathbf{a}\wedge(\mathbf{F}_1+\mathbf{F}_2+\ldots+\mathbf{F}_n)\\&=\mathbf{G}-\mathbf{a}\wedge\mathbf{F}.\end{aligned}\tag{24}$$

If $\mathbf{F}=0$ we say that the system reduces to a *couple* $\mathbf{G}$, which we see is independent of the centre of moments. This couple may be produced by two forces, $\mathbf{F}_1$ at $\mathbf{r}_1$ and $-\mathbf{F}_1$ at $\mathbf{r}_2$, such that $(\mathbf{r}_1-\mathbf{r}_2)\wedge\mathbf{F}_1=\mathbf{G}$. If $\mathbf{F}\neq 0$ the system is equivalent to a force $\mathbf{F}$ at O and a couple $\mathbf{G}$, or a force $\mathbf{F}$ at O' and a couple $\mathbf{G}'$.

Taking the scalar product with $\mathbf{F}$, we obtain from (24)

$$\mathbf{F}\cdot\mathbf{G}'=\mathbf{F}\cdot\mathbf{G}.$$

Unless $\mathbf{F}\cdot\mathbf{G}=0$, there can be no point O' such that $\mathbf{G}'=0$, and so the system cannot be reduced to a single force. Thus the concept of a resultant with a definite line of action which is useful in two dimensions (where the condition $\mathbf{F}\cdot\mathbf{G}=0$ is always satisfied) has little value in three dimensions. We may, however, look for a point O' at which $\mathbf{G}'$ is parallel to $\mathbf{F}$. If this is so $\mathbf{F}\wedge\mathbf{G}'=0$ and hence from (24)

$$\mathbf{F}\wedge\mathbf{G}-F^2\mathbf{a}+(\mathbf{F}\cdot\mathbf{a})\mathbf{F}=0.$$

Thus we certainly must have

$$\mathbf{a}=(\mathbf{F}\wedge\mathbf{G})/F^2+\lambda\mathbf{F}\tag{25}$$

where λ is a scalar. In this case

$$\mathbf{G}'=\mathbf{G}-\{(\mathbf{F}\wedge\mathbf{G})/F^2+\lambda\mathbf{F}\}\wedge\mathbf{F}=(\mathbf{F}\cdot\mathbf{G}/F^2)\mathbf{F},$$

and hence $\mathbf{G}'$ is parallel to $\mathbf{F}$ for any $\mathbf{a}$ on the line given by (25), with λ a variable parameter. This line is called the *central axis*. We see that the system of forces is equivalent to a force $\mathbf{F}$ acting along the central axis, together with a parallel couple $(\mathbf{F}\cdot\mathbf{G}/F^2)\mathbf{F}$, this constituting what is called a *wrench*.

Example 1. *The force* $\mathbf{F}=2\mathbf{i}-4\mathbf{j}-3\mathbf{k}$ *acts at the point*

$\mathbf{r}=\mathbf{i}-2\mathbf{j}+3\mathbf{k}$. *Find the moment of* $\mathbf{F}$ *about* O, *and also its moment about the line through* O *in the direction of the vector* $\mathbf{i}+\mathbf{j}+\mathbf{k}$.

The moment about O is $\mathbf{r}\wedge\mathbf{F}=(6+12)\mathbf{i}+(6+3)\mathbf{j}+(-4+4)\mathbf{k}=18\mathbf{i}+9\mathbf{j}$. The moment about the given line is

$$(1/\sqrt{3})(\mathbf{i}+\mathbf{j}+\mathbf{k})\cdot(18\mathbf{i}+9\mathbf{j})=(18+9)/\sqrt{3}=9\sqrt{3}.$$

Example 2. *Find the central axis and the equivalent wrench of the forces* $\mathbf{F}_1=3\mathbf{i}+2\mathbf{j}-\mathbf{k}$ *at* $\mathbf{r}_1=\mathbf{i}-2\mathbf{j}$ *and* $\mathbf{F}_2=\mathbf{i}-2\mathbf{j}+\mathbf{k}$ *at* $\mathbf{r}_2=\mathbf{j}-\mathbf{k}$.

$$\mathbf{F}=\mathbf{F}_1+\mathbf{F}_2=4\mathbf{i},\quad F=4.$$
$$\mathbf{G}=\mathbf{r}_1\wedge\mathbf{F}_1+\mathbf{r}_2\wedge\mathbf{F}_2=(2\mathbf{i}+\mathbf{j}+8\mathbf{k})+(-\mathbf{i}-\mathbf{j}-\mathbf{k})=\mathbf{i}+7\mathbf{k}.$$
$$\mathbf{F}\wedge\mathbf{G}=-28\mathbf{j},\quad \mathbf{F}\cdot\mathbf{G}=4.$$

The central axis is $\mathbf{r}=\lambda\mathbf{i}-\frac{7}{4}\mathbf{j}$.

The wrench has force $4\mathbf{i}$, couple $\mathbf{i}$.

EXERCISES ON CHAPTER ONE

1. If $\mathbf{a}=8\mathbf{i}-2\mathbf{j}+4\mathbf{k}$ and $\mathbf{b}=2\mathbf{i}+\mathbf{k}$, find the moduli and the direction cosines of $\mathbf{a}-\mathbf{b}$ and $\mathbf{a}\wedge\mathbf{b}$, and the angle between the directions of $\mathbf{a}$ and $\mathbf{b}$.

2. Prove by vector methods that
   (i) the diagonals of a parallelogram bisect one another,
   (ii) the perpendicular bisectors of the sides of a triangle are concurrent.

3. Find the equation of the plane through the points with position vectors $\mathbf{a}$, $\mathbf{b}$ and $\mathbf{c}$.

4. Find two vectors $\mathbf{a}$, $\mathbf{b}$ which are of equal magnitude, are mutually perpendicular, each have x-component 5, and are each perpendicular to $3\mathbf{j}+4\mathbf{k}$.

5. Prove that
   (i) $(\mathbf{a}\wedge\mathbf{b})\cdot(\mathbf{c}\wedge\mathbf{d})=(\mathbf{a}\cdot\mathbf{c})(\mathbf{b}\cdot\mathbf{d})-(\mathbf{a}\cdot\mathbf{d})(\mathbf{b}\cdot\mathbf{c})$,
   (ii) $[\mathbf{a}\wedge\mathbf{b},\ \mathbf{b}\wedge\mathbf{c},\ \mathbf{c}\wedge\mathbf{a}]=[\mathbf{a},\mathbf{b},\mathbf{c}]^2$.

6. Solve for $\mathbf{x}$ the equation $\mathbf{x}+\mathbf{x}\wedge\mathbf{a}=\mathbf{b}$.

7. A ship steaming north at 12 knots passes a fixed buoy at noon. A second ship steaming east at 16 knots passes the same buoy at 12.50 p.m. At what times are the ships closest together, and what is then their distance apart? [1 knot=1 sea mile per hour.]

8. Find the moment about the point $3\mathbf{i}+\mathbf{j}$ and the moment about the line $\mathbf{r}=(3-4\lambda)\mathbf{i}+\mathbf{j}+3\lambda\mathbf{k}$ of the force $7\mathbf{i}-\mathbf{j}+2\mathbf{k}$ acting at the point $6\mathbf{j}-\mathbf{k}$.

EXERCISES

9. Prove that the three forces at O represented by $\overrightarrow{OA}$, $\overrightarrow{OB}$ and $\overrightarrow{OC}$ are equivalent to the force $3\overrightarrow{OG}$ at O, where G is the centroid of the triangle ABC.

10. A rigid body is acted on by the force $\mathbf{F}_1 = \mathbf{i} - 3\mathbf{j} - 2\mathbf{k}$ at the point $-2\mathbf{i} + 9\mathbf{j}$ and by $\mathbf{F}_2 = 2\mathbf{i} + \mathbf{j} - 3\mathbf{k}$ at $-\mathbf{i} + y\mathbf{j} - \mathbf{k}$. For what value of y could the body be in equilibrium under the action of $\mathbf{F}_1$, $\mathbf{F}_2$ and a third force $\mathbf{F}_3$? Find $\mathbf{F}_3$ and its line of action in this case.

CHAPTER TWO
Dynamics of a Particle

2.1. THE NATURE OF DYNAMICS

Dynamics is the study of the motion of bodies. Like any other branch of applied mathematics it is essentially an experimental science, its purpose being to gain an understanding of phenomena in the physical world. The procedure is as follows. On the basis of preliminary observations certain basic laws, such as Newton's laws of motion, are stated in mathematical form. By standard mathematical techniques the consequences of these laws are then worked out, and in particular the behaviour is predicted in as wide as possible a range of specific physical situations which may be observed experimentally. If the results of the experiments agree with the predictions, this gives encouragement to continue the mathematical study of the equations, and confidence in applying the theoretical results. If the results disagree, the basic laws must be modified. In this way we may hope to increase progressively our understanding of the relationship between the various parts of physical experience.

The value of a law thus depends on the number of physically useful deductions which can be made from it. The 'truth' of the law is irrelevant; indeed the expression is meaningless from the scientific point of view. All experiments contain some inaccuracies of measurement, and effects other than those allowed for in the calculations cannot be totally excluded, so perfect agreement cannot be looked for. Nor does it matter if in certain circumstances a law ceases to apply, provided that we appreciate its limitations. At speeds comparable with the speed of light, Newton's laws of motion are no longer accurate and have to be replaced by relativistic laws. However, at lower speeds to include the relativistic terms in the equations is a positive disadvantage, since it greatly complicates the mathematics without giving any further useful information. On a more mundane level, to find the trajectory of a thrown cricket ball a simple calculation as in Example 3 of section 2.2 is often sufficient. For a better approximation we may allow for air resistance, assuming a plausible resistance law. But still other effects could be included, such as the rotation of the earth, the spin of the ball, and the variation of the wind in time and space. The whole problem would then be extremely complicated, and the final answer, if one was ever obtained, would probably be no more useful than the earlier ones. The recognition of what agencies are important and what are not is essential in any piece of applied mathematics.

The aim of the remainder of this book is to develop dynamics in a precise and logical way, making it clear at each stage what assumptions are being made and to what extent new results are just mathematical deductions from previous ones. Examples will be given to illustrate the dynamical principles, and to call attention to the wide range of physical situations which may be studied, though for reasons of space many interesting problems must be left unmentioned.

2.2. NEWTON'S LAWS

Newton's laws of motion may be stated as follows:

1. *A particle continues in a state of rest, or of uniform motion in a straight line, unless acted on by a force.*

2. *The acceleration of a particle produced by a force is proportional to the force and in the direction of the force.*

3. *If two particles interact, the action and reaction are equal and opposite.*

We shall not have time here to discuss the processes by which Newton was led to formulate these laws. Although so widely used, acceptance of the laws does of course rest solely on the experimental verification of their consequences.

The meaning to be attached to the word *particle* in the laws must be studied first. A particle is often said to be a point mass with no spatial extent. Atomic nuclei and electrons might be thought to be particles of this type. But Newton's laws are not intended to apply to such small-scale phenomena; usually quantum mechanics must be used instead. In classical mechanics the smallest pieces of matter we need to consider contain enormous numbers of atoms, and on this scale we can ignore the atomic structure and think of matter as continuous.

Accordingly, we define a particle to be a material body whose dimensions, though not zero, are sufficiently small for the internal structure of the body to be unimportant. The actual size permissible depends on the particular physical problem. Thus the earth may be treated as a single particle for the discussion of its movement round the sun, but a grain of sand cannot be treated as one in a study of the formation of a sand dune. For our calculations, the essential feature of a particle is that its position is sufficiently described by a single vector $\mathbf{r}$, the position vector from some origin.

This concept of a particle in Newton's laws will not be self-consistent unless we can prove that a number of particles, fixed together and of small extent, behave like a single particle, for we do not want to have to specify exactly how small our particles must be. The required proofs are obtained in section 3.1 and section 3.3.

The laws of motion for a particle would be of limited use if we could not deduce from them the laws applicable to larger bodies and to systems of several bodies. Our definition of a particle enables us to deal with extended systems by considering them as assemblages of suitably interconnected smaller bodies, all sufficiently small to be treated as single particles.

Newton's second law, of which the first is a special case,

says that the acceleration $\ddot{\mathbf{r}}$ is proportional to the applied force $\mathbf{F}$, or

$$\mathbf{F} = m\ddot{\mathbf{r}}. \tag{26}$$

The constant of proportionality m is called the *mass* of the particle. Now $\ddot{\mathbf{r}}$ is a vector since $\mathbf{r}$ is, from section 1.6, and hence $\mathbf{F}$ is a vector, being equal to the product of a scalar and a vector. Thus Newton's second law implies that force is a vector, and from this the laws for the combination of forces follow as discussed in section 1.8. Equation (26) enables us not only to define, but also to measure, forces and masses. If we apply a given force $\mathbf{F}$ (say by attaching a spring stretched by a definite amount) successively to each of a series of particles of masses $m_1, m_2, \ldots$, and measure the accelerations $\ddot{\mathbf{r}}_1, \ddot{\mathbf{r}}_2, \ldots$, then since the accelerations are given to be inversely proportional to the masses, we can assign numbers to specify the masses, having arbitrarily chosen one of the particles to have unit mass. And if we apply forces in turn to the standard particle, of unit mass, the forces are equal to the accelerations produced.

This definition of mass makes no reference whatever to the *weight* of a particle, which is the magnitude of the gravitational force which acts upon it. It is an additional piece of observational experience that, in the absence of air resistance and other disturbing forces, all particles fall under gravity with the same constant acceleration $\mathbf{g}$. The weight of a particle of mass m is hence mg, from (26), and a comparison of weights therefore also gives a comparison of masses. It is important to realize that Newton's laws of motion would provide a completely satisfactory basis for dynamics if the gravitational force were *not* proportional to the mass, or indeed did not exist at all. Weight varies significantly even between one point on the earth's surface and another, while mass is constant (except at speeds comparable with the speed of light). It is mass which should always be thought of as the fundamental property of matter for dynamical purposes.

In the British system the units of length, mass and time are the foot, pound and second, all chosen quite arbitrarily. The unit of force, the poundal, gives a mass of 1 lb. an acceleration of 1 ft./sec.^{2}. Since at the earth's surface g is approximately 32·2 ft./sec.^{2}, the weight of a mass of 1 lb. is 32·2 poundals; this force is sometimes called 1 pound weight. In the c.g.s. system the units of length, mass and time are the centimetre, gram and second, and the corresponding unit of force is the dyne. Since g is approximately 981 cm./sec.^{2}, a force of 1 gm.wt. is equal to 981 dynes.

Many simple problems concerned with the motion of a particle can be solved by a direct application of Newton's second law.

Example 1. *A body moves along the* x*-axis with constant acceleration* a, *starting at* x=0 *at time* t=0 *with speed* u. *Find the speed* v *in terms of* x *and in terms of* t, *and find* x *in terms of* t.

The acceleration $f=dv/dt=a$. Integrating with respect to t,

$$v=at+\text{constant}=u+at.$$

Now $v=dx/dt$, and integrating again with respect to t,

$$x=ut+\tfrac{1}{2}at^2.$$

No constant appears since $x=0$ when $t=0$. To find v as a function of x, we may eliminate t between the last two equations. Alternatively, we may write

$$f=\frac{dv}{dt}=\frac{dv}{dx}\cdot\frac{dx}{dt}=v\frac{dv}{dx}.$$

This formula is frequently in use, and should be remembered. Hence $v\,dv/dx=a$, and integrating with respect to x,

$$\tfrac{1}{2}v^2=ax+\text{constant},\quad v^2=u^2+2ax.$$

All the required formulae have now been found. It must be emphasized that they hold only for motion with constant acceleration.

Example 2. *A body falls from rest under the influence of gravity and air resistance proportional to the square of its speed. Find how the speed of the body depends on the distance fallen. (This resistance law is in good agreement with experience.)*

Let x be the distance measured downwards from the starting point. Equation (26) gives

$$mf = mg - kv^2,$$

where m is the mass of the body, f its acceleration, v its speed and k is a constant. If $v^2 < mg/k$, $f > 0$ and the body accelerates; if $v^2 > mg/k$, $f < 0$ and the body decelerates. In either case the speed approaches $V = (mg/k)^{\frac{1}{2}}$, which is known as the *terminal speed*. We may write the equation as

$$mf = k(V^2 - v^2).$$

We wish to find v as a function of x, and hence we write $f = v\, dv/dx$.

Integrating the equation, we then have

$$\frac{k}{m}x = \int \frac{v\, dv}{V^2 - v^2} = -\tfrac{1}{2}\log(V^2 - v^2) + \log V,$$

since $v = 0$ when $x = 0$. Taking the exponential of this, we obtain

$$v^2 = V^2\left\{1 - e^{-2kx/m}\right\} = \frac{mg}{k}\left\{1 - e^{-2kx/m}\right\}.$$

This gives the speed of the body throughout the fall.

If the body is projected upwards, the resistance acts downwards initially, and so the equation which governs the motion until the speed falls to zero is $mf = mg + kv^2$, in the same co-ordinates as before. The upward and downward parts of the motion must therefore be treated separately.

Example 3. *Find the path of a projectile which starts from the origin with velocity* **u** *at time* t=0, *neglecting air resistance.*

The equation of motion is

$$\ddot{\mathbf{r}} = -g\mathbf{k},$$

where $\mathbf{k}$ is a unit vector in the z-direction, vertically upwards. Integrating twice, we have first

$$\dot{\mathbf{r}}=\mathbf{u}-gt\mathbf{k},$$

since $\dot{\mathbf{r}}=\mathbf{u}$ when $t=0$, and then

$$\mathbf{r}=\mathbf{u}t-\tfrac{1}{2}gt^2\mathbf{k},$$

since $\mathbf{r}=0$ when $t=0$. If the direction of projection is at an elevation α above the horizontal in the xz-plane, the components of this equation give

$$x=ut\cos\alpha,\ y=0,\ z=ut\sin\alpha-\tfrac{1}{2}gt^2.$$

To find the path, we eliminate t and obtain

$$z=x\tan\alpha-\frac{gx^2}{2u^2}\sec^2\alpha.$$

This is a parabola, with its axis vertical. Other properties of the motion are readily deduced. Thus the horizontal range, found by putting $z=0$ with $x\neq 0$, is $(u^2/g)\sin 2\alpha$; this range has its greatest value u^2/g when $\alpha=\tfrac{1}{4}\pi$.

Example 4. *Find the extent of the region which can be reached by the projectile of Example* 3, *if* u *is kept constant but the angle of projection is varied.*

The equation for the path can be written as

$$\tan^2\alpha-\frac{2u^2}{gx}\tan\alpha+\left(1+\frac{2u^2z}{gx^2}\right)=0.$$

since $\sec^2\alpha=1+\tan^2\alpha$. The roots, if any, of this quadratic equation for $\tan\alpha$ give paths which pass through the point (x, z). There are real roots only if

$$\frac{u^4}{g^2x^2}\geqslant 1+\frac{2u^2z}{gx^2},\ \text{or}\ z\leqslant\frac{u^2}{2g}-\frac{gx^2}{2u^2}.$$

In the case of equality, the point (x, z) lies on a parabola with its axis vertical and vertex uppermost, the *parabola of safety*. Only points on and below this parabola can be reached by the projectile.

2.3. MOMENTUM, ANGULAR MOMENTUM, ENERGY

If we define a new vector, the *momentum* $\mathbf{M}=m\dot{\mathbf{r}}=m\mathbf{v}$, where $\mathbf{v}$ is the particle's velocity, equation (26) can be written as

$$\mathbf{F}=\frac{d\mathbf{M}}{dt}, \tag{27}$$

and thus the force applied is equal to the rate of change of the momentum. In particular if $\mathbf{F}=0$ throughout the motion, $\mathbf{M}$=constant. This is the principle of the conservation of momentum for a particle. Even if $\mathbf{F}$ is not zero, one of its components may be. In this case the corresponding component of $\mathbf{M}$ remains constant. Thus in Example 3 of section 2.2, $M_x=mu\cos\alpha$ is constant.

We may also take the vector product of (26) with $\mathbf{r}$, obtaining $\mathbf{r}\wedge\mathbf{F}=\mathbf{r}\wedge m\ddot{\mathbf{r}}$. Now $\mathbf{r}\wedge\mathbf{F}=\mathbf{G}$, the moment of $\mathbf{F}$ about O as defined in section 1.8. We introduce another new vector, the *angular momentum* $\mathbf{H}=\mathbf{r}\wedge m\dot{\mathbf{r}}=\mathbf{r}\wedge m\mathbf{v}$, which we see is equal to the moment about O of the momentum. Since m is constant, $d\mathbf{H}/dt=\mathbf{r}\wedge m\ddot{\mathbf{r}}+\dot{\mathbf{r}}\wedge m\dot{\mathbf{r}}$, and the last term is zero. Hence our equation may be written as

$$\mathbf{G}=\frac{d\mathbf{H}}{dt}. \tag{28}$$

Thus the moment about O of the applied force is equal to the rate of change of the angular momentum about O. If $\mathbf{G}=0$, $\mathbf{H}$=constant, this being the principle of the conservation of angular momentum.

Just as for the moment of a force, discussed in section 1.8, we may define the component of $\mathbf{H}$ in any direction to be the angular momentum about a line through O in that direction, in agreement with the usual definition in two-dimensional dynamics.

These laws of momentum and angular momentum are direct consequences of Newton's second law. We shall see in Chapter Three that the concepts of momentum and angular momentum are of great use in studying the motion of a system of particles.

If we multiply (26) scalarly by $\dot{\mathbf{r}}$, we obtain

$$\mathbf{F} \cdot \dot{\mathbf{r}} = \frac{dT}{dt} \tag{29}$$

where $T = \frac{1}{2}m\dot{\mathbf{r}}^2 = \frac{1}{2}mv^2$ is a scalar quantity, the *kinetic energy*. Now $\mathbf{F} \cdot \delta\mathbf{r}$ is the work done (in time δt) by the force $\mathbf{F}$ in the displacement $\delta\mathbf{r}$, as discussed in section 1.3, and hence the rate of working of $\mathbf{F}$ is $\mathbf{F} \cdot \dot{\mathbf{r}}$. This is equal to the magnitude of $\mathbf{F}$ multiplied by the component of the velocity in the direction of $\mathbf{F}$. Equation (29) shows that the rate of working of the applied force on a particle is equal to the rate of change of the kinetic energy.

For special forms of $\mathbf{F}$ there are advantages in using (29) rather than the original equation (26). In many physical situations the force $\mathbf{F}$ is a function of position only (not depending on the particle's velocity or the time), and does no net work when its point of application moves round any closed curve. Such a force is said to be *conservative*.

We can show that in this case the work done in a displacement depends only on the initial and final positions, and not on the path traversed. Consider the two paths, ABP, ACP from A to P shown in Fig. 12. The work for the closed path $ABPCA$ is known to be zero, and thus the sum of the amount of work done on ABP and PCA is zero.

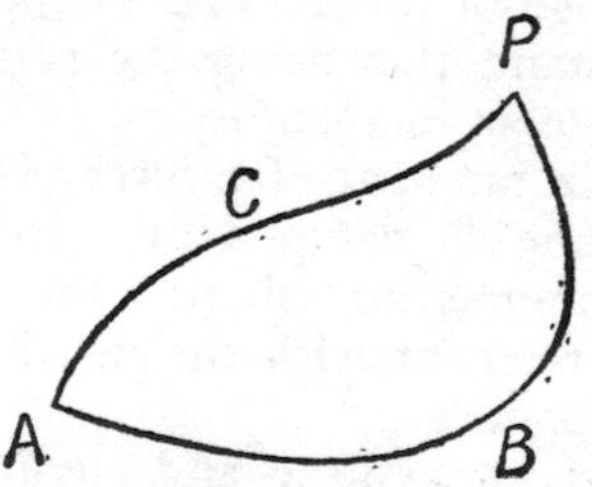

Figure 12

But the work on *PCA* is minus the work on *ACP*, since $\mathbf{F}$ depends only on position and for each element $\delta\mathbf{r}$ of the path the sign is changed. This shows that the work done on *ABP* is equal to the work done on *ACP*, and since these paths are quite general ones from A to P the result follows.

We may now define the *potential energy* $V(\mathbf{r})$ of the conservative force $\mathbf{F}(\mathbf{r})$ as the work done by $\mathbf{F}$ when its point of application is moved to a standard position A, no specification of the path being necessary. This is equal to minus the work done in a displacement from A to the present position. In a small displacement $\delta\mathbf{r}$, the change in V is $\delta V = -\mathbf{F} \cdot \delta\mathbf{r}$, since this is the extra amount of work done by $\mathbf{F}$ in returning to its starting position, and thence to A. Dividing by the time δt in which the displacement takes place and taking the limit as $\delta t \to 0$, we see that $dV/dt = -\mathbf{F} \cdot \dot{\mathbf{r}}$, and hence from (29)

$$\frac{dT}{dt} + \frac{dV}{dt} = 0,$$

and on integration

$$T + V = \text{constant}. \tag{30}$$

This is the energy equation. The fact that V and T appear on an equal footing in this equation justifies our calling them both by the name of energy.

The point A used in defining the potential energy may be chosen arbitrarily. If the point A' is used instead of A, $V(\mathbf{r})$ is changed by a constant amount, equal to the work done by $\mathbf{F}$ in a displacement from A to A', and the only effect of this is to alter the value of the constant in equation (30).

An advantage of using the energy equation when available is that it is an integrated form of the equation of motion, involving only the velocity $\dot{\mathbf{r}}$ and not the acceleration $\ddot{\mathbf{r}}$. However, it is only a single equation, while the components of the vector equation for the momentum (or the angular momentum) provide three equations to determine the dependence on time of the three components of $\mathbf{r}$. Unless the motion can be specified by a single co-ordinate, the energy equation alone is insufficient and suitable other equations must also be taken into account.

Example 1. *The position vector at time* t *of a particle of mass* m *is* $\bar{\mathbf{r}}=\mathrm{a}\cos \mathrm{nt}\,\mathbf{i}+\mathrm{b}\sin \mathrm{nt}\,\mathbf{j}$. *Show that the angular momentum about* O *is constant, and find the force and moment on the particle.*

We have

$$\dot{\mathbf{r}}=-an\sin nt\,\mathbf{i}+bn\cos nt\,\mathbf{j},$$

$$\ddot{\mathbf{r}}=-an^2\cos nt\,\mathbf{i}-bn^2\sin nt\,\mathbf{j}=-n^2\mathbf{r}.$$

Hence

$$\mathbf{H}=\mathbf{r}\wedge m\dot{\mathbf{r}}=mabn(\cos^2 nt+\sin^2 nt)\mathbf{k}=mabn\mathbf{k},$$

and so is constant. Also

$$\mathbf{F}=m\ddot{\mathbf{r}}=-mn^2\mathbf{r},\ \mathbf{G}=\mathbf{r}\wedge\mathbf{F}=0,$$

which is as required for the proved conservation of the angular momentum.

Example 2. *Calculate the potential energy of the forces* $-\mathrm{mg}\mathbf{k}$ *and* $-(\mu/\mathrm{r}^2)\hat{\mathbf{r}}$.

If $\mathbf{F}=-mg\mathbf{k}$, work is done only by the component of displacement in the z-direction, and hence

$$V=-\int(-mg)dz=mgz.$$

This is the gravitational potential of a mass m near the earth's surface. The height $z=0$ may be chosen arbitrarily.

If $\mathbf{F}=-(\mu/r^2)\hat{\mathbf{r}}$, work is done only by the radial component of displacement, and hence

$$V=-\int(-\mu/r^2)dr=-\mu/r.$$

This is the potential energy due to an inverse square law of attraction towards the origin.

Example 3. *A particle of mass* m *moves along the* x*-axis, controlled by a light elastic spring which produces a restoring force* $-\mathrm{kx}\mathbf{i}$. *Find the potential energy stored in the spring, and calculate the position of the particle at time* t *if it is released from rest at* x=a *at time* t=0.

The potential energy $V=-\int(-kx)dx=\frac{1}{2}kx^2$. The energy equation for the motion is

$$\tfrac{1}{2}m\dot{x}^2+\tfrac{1}{2}kx^2=\text{constant}=\tfrac{1}{2}ka^2,$$

since initially $x=a$, $\dot{x}=0$. Hence

$$\dot{x}=\left\{\frac{k}{m}(a^2-x^2)\right\}^{\frac{1}{2}},\quad \left(\frac{k}{m}\right)^{\frac{1}{2}}t=\int\frac{dx}{(a^2-x^2)^{\frac{1}{2}}}=\sin^{-1}\left(\frac{x}{a}\right)-\tfrac{1}{2}\pi,$$

since $x=a$ when $t=0$. The required position is thus given by

$$x=a\cos\{(k/m)^{\frac{1}{2}}t\}.$$

This type of motion, known as *simple harmonic motion*, is of very frequent occurrence. The whole motion is periodic in time, with period $2\pi(m/k)^{\frac{1}{2}}$.

The energy equation could have been arrived at by integrating (after multiplication by $\dot{x}$) the equation of motion $m\ddot{x}=-kx$, or

$$\ddot{x}+(k/m)x=0.$$

This may be considered as the basic form of equation for simple harmonic motion, with (k/m) any positive constant.

2.4. ACCELERATING FRAMES OF REFERENCE

We have as yet taken no particular account of the frame of reference (i.e. the origin and the direction of the axes) used to describe the motion. The position vector $\mathbf{r}$, the velocity $\dot{\mathbf{r}}$ and the acceleration $\ddot{\mathbf{r}}$ which we measure will be affected by motion of the origin, but the forces will not be, and so Newton's second law cannot possibly hold in *all* frames of reference, unless we postulate the presence of fictitious forces to make up any discrepancy. The idea of fictitious forces is not to be immediately dismissed, as we shall see below.

At one time the idea of absolute motion was thought to have validity, but now it is generally accepted that all motion is relative, and there is no question of our being able to say categorically that any point is at rest, or even moving with uniform velocity. In any case we often wish to take as origin a point on the surface of the earth, which is obviously moving. Such a point rotates about the earth's axis once in a day, travels round the sun in a year, and follows the sun in its path through the galaxy, all at speeds far in excess of those we wish to be able to measure.

Though it may seem illogical to start from laws expressed in a form which appears to require that absolute motion exists, our best course is to postulate that we have a basic *Newtonian frame of reference*, with origin O, in which Newton's laws apply. Then we can consider the equations of motion in frames of reference which move relative to the basic frame, and can estimate the effects of such movements. Eventually we may hope to put the laws into such a form that the original frame of reference no longer appears explicitly. Results of this nature will be obtained in section 3.2.

Let us suppose that instead of O, we take as origin the point O', where $\overrightarrow{OO'}=\mathbf{a}$ as shown in Fig. 13. Relative to O'

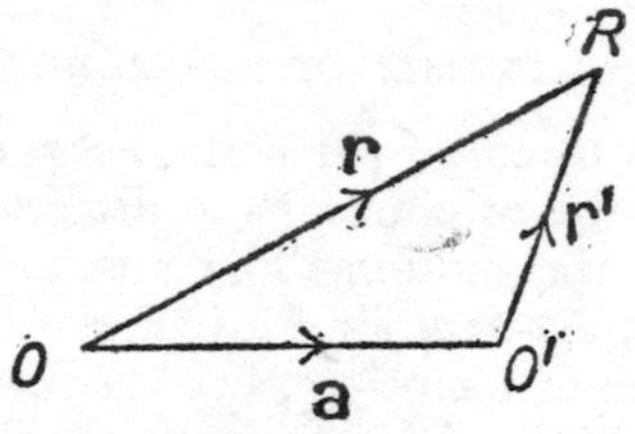

Figure 13

the position vector of a particle at R is $\mathbf{r}'=\mathbf{r}-\mathbf{a}$. The equation of motion (26) shows that

$$\mathbf{F}=m\ddot{\mathbf{r}}=m\ddot{\mathbf{r}}'+m\ddot{\mathbf{a}},$$

$$m\ddot{\mathbf{r}}'=\mathbf{F}-m\ddot{\mathbf{a}}. \qquad (31)$$

Comparing this with equation (26) we see that in the new

frame of reference there is a new apparent force $-\ddot{\mathbf{a}}$ per unit mass, an *inertial force* acting in the opposite direction to the acceleration of O'. If $\ddot{\mathbf{a}}=0$, i.e. $\dot{\mathbf{a}}=$constant, no extra force arises, and thus all frames of reference which have constant velocity relative to a Newtonian frame are themselves Newtonian frames.

A familiar example of this inertial force is seen in a lift accelerating upwards, when an occupant is conscious of an extra downward force, though he remains at rest relative to his surroundings. On the assumption that inertial and gravitational masses are equal, the observed terrestrial force of gravity could be reproduced exactly in empty space by giving the frame of reference an acceleration $-\mathbf{g}$. Einstein's general theory of relativity postulates that accelerations and gravitational forces are essentially equivalent; on this basis a gravitational force can equally well be thought of as an inertial force, and hence weight *must* be proportional to mass and does not require any separate assumption.

Example 1. *A train is moving along a straight horizontal track with uniform acceleration* f. *Calculate the motion relative to the train of a parcel dropped by a passenger.*

The parcel is subject to its weight mg downwards and to an inertial force $-mf$ in the direction of the train's motion; hence the total force is of magnitude $m(g^2+f^2)^{\frac{1}{2}}$ and inclined at an angle $\tan^{-1}(f/g)$ to the vertical. Since the force is constant, the motion is exactly as treated in Example 1 of section 2.2. The parcel falls in a straight line making an angle $\tan^{-1}(f/g)$ with the vertical, its speed and distance travelled being given by the formulae of Example 1 of section 2.2 with a replaced by $(g^2+f^2)^{\frac{1}{2}}$, and with $u=0$ in this particular problem.

2.5. ROTATING FRAMES OF REFERENCE

Suppose that the origin O is fixed in a Newtonian frame, but that the axes with respect to which position and displacement are measured rotate. At time t, let any vector quantity $\mathbf{a}(t)$ be given in terms of its components by

$$\mathbf{a}=a_x\mathbf{i}+a_y\mathbf{j}+a_z\mathbf{k},$$

where $\mathbf{i}$, $\mathbf{j}$, $\mathbf{k}$ are unit vectors in the co-ordinate directions. If the axes rotate, we can consider that these unit vectors are embedded in a rigid body which rotates about O with angular velocity $\boldsymbol{\omega}$. From equation (19), the rates of change with respect to fixed axes are

$$d\mathbf{i}/dt=\boldsymbol{\omega}\wedge\mathbf{i},\ d\mathbf{j}/dt=\boldsymbol{\omega}\wedge\mathbf{j},\ d\mathbf{k}/dt=\boldsymbol{\omega}\wedge\mathbf{k},$$

and hence, by the rules for vector differentiation, the rate of change of $\mathbf{a}(t)$ is

$$\begin{aligned}\frac{d\mathbf{a}}{dt}&=\frac{da_x}{dt}\mathbf{i}+\frac{da_y}{dt}\mathbf{j}+\frac{da_z}{dt}\mathbf{k}+a_x\frac{d\mathbf{i}}{dt}+a_y\frac{d\mathbf{j}}{dt}+a_z\frac{d\mathbf{k}}{dt}\\&=\frac{\partial\mathbf{a}}{\partial t}+\boldsymbol{\omega}\wedge\mathbf{a}.\end{aligned}\qquad(32)$$

Here $\partial\mathbf{a}/\partial t$ is written for the rate of change of $\mathbf{a}(t)$ in the moving frame, with the motion of the axes ignored. If the components of $\mathbf{a}$ remain constant, so that $\mathbf{a}$ is represented by a position vector carried round with the axes, $\partial\mathbf{a}/\partial t=0$. Equation (32) is of fundamental importance, and has many applications.

In the case in which $\mathbf{a}=\mathbf{r}$, the position vector, we shall write $\partial\mathbf{r}/\partial t=\dot{\mathbf{r}}$, the 'apparent' velocity in the moving frame, and shall write $d\mathbf{r}/dt=\mathbf{v}$, the 'true' velocity in the basic Newtonian frame. Then

$$\mathbf{v}=\dot{\mathbf{r}}+\boldsymbol{\omega}\wedge\mathbf{r}.\qquad(33)$$

Likewise, the 'true' acceleration $\mathbf{f}$ is given by

$$\begin{aligned}\mathbf{f}&=d\mathbf{v}/dt\\&=\partial\mathbf{v}/\partial t+\boldsymbol{\omega}\wedge\mathbf{v}\\&=\ddot{\mathbf{r}}+\dot{\boldsymbol{\omega}}\wedge\mathbf{r}+2\omega\wedge\dot{\mathbf{r}}+\omega\wedge(\boldsymbol{\omega}\wedge\mathbf{r}),\end{aligned}\qquad(34)$$

from (33). For a particle of mass m acted on by a force $\mathbf{F}$, the observed acceleration $\ddot{\mathbf{r}}$ in the moving frame is given by (26) and (34) in the form

$$m\ddot{\mathbf{r}}=\mathbf{F}-m\dot{\boldsymbol{\omega}}\wedge\mathbf{r}-2m\omega\wedge\dot{\mathbf{r}}-m\omega\wedge(\boldsymbol{\omega}\wedge\mathbf{r}).\qquad(35)$$

Written in this way, the last three terms may be considered as apparent forces which arise owing to the rotation of the axes. The term $-m\dot{\boldsymbol{\omega}} \wedge \mathbf{r}$ is an *inertial force*, exactly analogous to that found in (31), and is proportional to the angular acceleration of the axes. In the last term, $\boldsymbol{\omega} \wedge \mathbf{r}$ has magnitude ωp, where p is the distance of the particle from the axis of rotation, and is perpendicular to both $\boldsymbol{\omega}$ and $\mathbf{r}$; consequently $-m\boldsymbol{\omega} \wedge (\boldsymbol{\omega} \wedge \mathbf{r})$ acts directly away from the axis and has magnitude $m\omega^2 p$. It is the familiar *centrifugal force*. The remaining term $-2m\boldsymbol{\omega} \wedge \dot{\mathbf{r}}$ arises only when the particle is in motion relative to the moving axes, and gives what is called the *Coriolis force*. It acts in a direction perpendicular to both $\boldsymbol{\omega}$ and the apparent velocity $\dot{\mathbf{r}}$. If the axes have a translational motion as well as rotation, the inertial force of equation (31) will be observed in addition.

We can now examine the effect of the rotation of the earth on the dynamics of motion near its surface. Choose an origin O on the axis of rotation of the earth near its centre, and use axes fixed in the earth. Consider a particle of mass m with position vector $\mathbf{r}$, acted on by an applied force $\mathbf{F}'$ as well as by the gravitational force $m\mathbf{g}'$. Since the angular velocity $\boldsymbol{\Omega}$ of the earth is constant, (35) gives

$$m\ddot{\mathbf{r}} = \mathbf{F}' + m\{\mathbf{g}' - \boldsymbol{\Omega} \wedge (\boldsymbol{\Omega} \wedge \mathbf{r})\} - 2m\boldsymbol{\Omega} \wedge \dot{\mathbf{r}}. \qquad (36)$$

The reason for grouping together the gravitational and centrifugal terms is that both are functions of position only; indeed in their dynamical effects they are inseparable. We customarily define $\mathbf{g}$ to be the acceleration of a body falling freely from rest, as measured in axes fixed in the earth. In this case $\mathbf{F}'=0$, $\dot{\mathbf{r}}=0$, $\ddot{\mathbf{r}}=\mathbf{g}$, and hence from (36)

$$\mathbf{g} = \mathbf{g}' - \boldsymbol{\Omega} \wedge (\boldsymbol{\Omega} \wedge \mathbf{r}). \qquad (37)$$

The difference between the observed gravity $\mathbf{g}$ and the true gravity $\mathbf{g}'$ is quite appreciable, as the following example shows.

Example 1. *Calculate the difference between* $\mathbf{g}$ *and* $\mathbf{g}'$ *at a point on the equator.* [*The radius of the earth is* 3,960 *miles.*]

If the earth rotated on its axis once in twenty-four hours, we should have $\Omega=2\pi/(24\times60\times60)=0{\cdot}0000727$ radians/sec. Since the earth goes round the sun once in a year, it in fact rotates approximately 366 times in 365 days, and hence $\Omega=0{\cdot}0000727\times366/365=0{\cdot}0000729$ radians/sec.

On the equator $\mathbf{\Omega}\wedge(\mathbf{\Omega}\wedge\mathbf{r})=-\Omega^2\mathbf{r}$ by (15) since $\mathbf{r}$ is perpendicular to $\mathbf{\Omega}$, and so $\mathbf{g}$ and $\mathbf{g}'$ are parallel and $g'-g=(0{\cdot}0000729)^2\times3{,}960\times5{,}280=0{\cdot}111$ ft./sec.^{2}.

This is of the same order of magnitude as the variation of g' between the equator and the poles due to the earth being an oblate spheroid and not a perfect sphere.

Equation (36) may now be written as

$$m\ddot{\mathbf{r}}=\mathbf{F}'+m\mathbf{g}-2m\mathbf{\Omega}\wedge\dot{\mathbf{r}}. \tag{38}$$

This equation is unaffected if we transfer the origin to O' on the earth's surface, a fixed point relative to the rotating

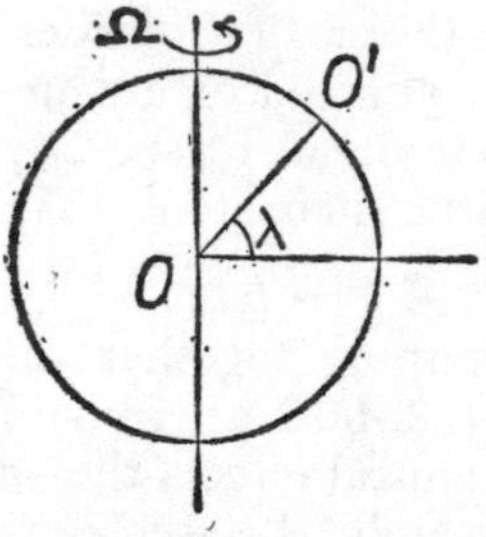

Figure 14

axes. Thus the only extra force which need be considered for the motion near a point on the earth's surface is the Coriolis force. This force has a governing influence on the winds over the earth, and it also has directly observable dynamical effects, as the following example shows.

Example 2. *A shell is fired with a horizontal velocity of* 2,000 *ft./sec. Estimate the deflection due to the Coriolis force at* *a range of* 10 *miles in latitude* 50° *N., neglecting air resistance.*

It is the direction of $\mathbf{g}$ which defines the vertical (and *not* the direction of $\mathbf{g}'$ or the radius from O' to the centre of the earth) and so $\mathbf{g}$ has no horizontal component. Take horizontal axes x in the direction of firing and y perpendicular to it. Since $\dot{\mathbf{r}}=\dot{x}\mathbf{i}+\dot{y}\mathbf{j}$, only the vertical component $\Omega \sin \lambda$ of $\boldsymbol{\Omega}$ (where λ is the latitude, shown in Fig. 14, equal to the complement of the angle between the vertical and the earth's axis) affects the equations for the horizontal motion, which from (38) are

$$\ddot{x}=2\Omega\dot{y} \sin \lambda, \; \ddot{y}= - 2\Omega\dot{x} \sin \lambda.$$

These equations show that in the northern hemisphere, where $\sin \lambda>0$, the deflection due to the Coriolis force is to the right, while in the southern hemisphere, where $\sin \lambda<0$, it is to the left. The reason for this is most easily understood physically for a shell fired to the north. In the northern hemisphere the shell passes into regions where the ground's speed to the east due to the earth's rotation is less, and so the shell deviates to the east. In the southern hemisphere the converse holds, and the shell deviates to the west.

Our equations may be integrated to give

$$\dot{x}- u=2\Omega y \sin \lambda, \; \dot{y}= - 2\Omega x \sin \lambda,$$

where u is the horizontal velocity of projection, provided that the change in latitude is insignificant. The Coriolis force is relatively small, so from the first equation $x=ut$, approximately, and using this we can integrate the second equation to obtain $y= - \Omega u t^2 \sin \lambda$.

In the present example $\Omega=0{\cdot}0000729$ radians/sec., the time of flight $t=52800/2000$ sec., $u=2000$ ft./sec., $\sin \lambda=0{\cdot}7660$. Inserting these values we obtain $y=78$ ft. This is of amply sufficient size to have to be allowed for in accurate gunnery.

2.6. POLAR CO-ORDINATES

We may use (33) and (34) to derive the components of velocity and acceleration in *plane polar co-ordinates* (r, θ).

Let the axes rotate so that **i** is always in the direction of the radius vector and **j** perpendicular to it in the plane of motion, as shown in Fig. 15. Then $\mathbf{r}=r\mathbf{i}$, $\dot{\mathbf{r}}=\dot{r}\mathbf{i}$, $\ddot{\mathbf{r}}=\ddot{r}\mathbf{i}$. The axes rotate about the normal to the plane of motion through the origin with angular velocity $\boldsymbol{\omega}=\dot{\theta}\mathbf{k}$, and therefore $\boldsymbol{\omega}\wedge\mathbf{r}=r\dot{\theta}\mathbf{j}$. Equations (33) and (34) now show that

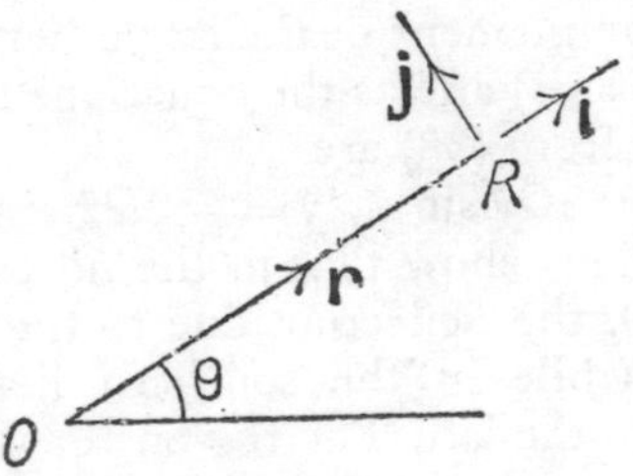

Figure 15

$$\mathbf{v}=\dot{r}\mathbf{i}+r\dot{\theta}\mathbf{j},\ \mathbf{f}=(\ddot{r}-r\dot{\theta}^2)\mathbf{i}+(r\ddot{\theta}+2\dot{r}\dot{\theta})\mathbf{j}. \tag{39}$$

The coefficients of **i** and **j** give the radial and transverse components respectively.

We can also deduce the components of acceleration for plane motion in terms of *intrinsic co-ordinates*, i.e. the distance s along the path and the angle ψ which the direction of motion makes with a fixed direction. Take unit vectors **i** and **j** parallel and perpendicular to the direction of motion as shown in Fig. 16. Then $\mathbf{v}=v\mathbf{i}$ and

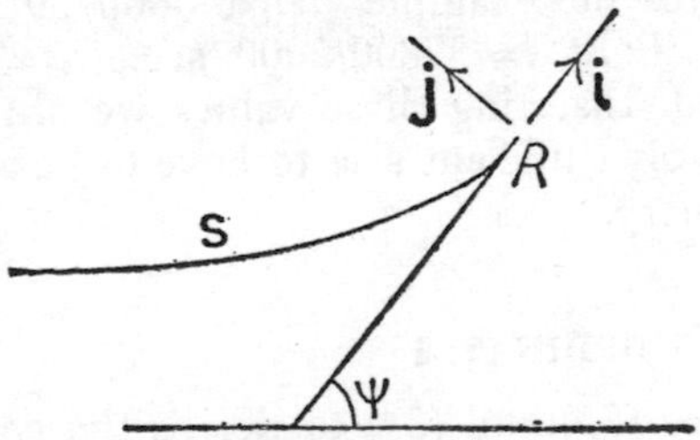

Figure 16

$$\boldsymbol{\omega}=\frac{d\psi}{dt}\mathbf{k}=\frac{d\psi}{ds}\cdot\frac{ds}{dt}\mathbf{k}=\frac{v}{\rho}\mathbf{k}$$

where $\rho=ds/d\psi$ is the radius of curvature of the path. From (32)

$$\mathbf{f}=\dot{\mathbf{v}}+\boldsymbol{\omega}\wedge\mathbf{v}=\dot{v}\mathbf{i}+(v^2/\rho)\mathbf{j}. \tag{40}$$

Thus the acceleration components are $\dot{v}$ parallel to the path and v^2/ρ perpendicular to it.

Example 1. *A particle of mass* m *hangs by a light string of length* l *from a fixed point* O. *Examine the motion in a vertical plane through* O.

Let the string make an angle θ with the downward vertical, as shown in Fig. 17, and let the tension in the string be T. From (39) the equations of motion in the radial and transverse directions are

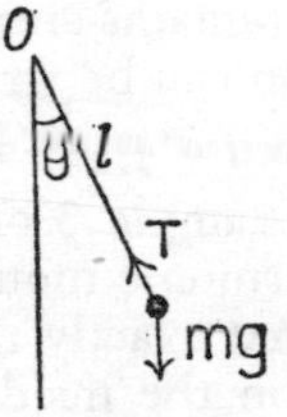

Figure 17

$$mg\cos\theta-T=-ml\dot{\theta}^2,\quad -mg\sin\theta=ml\ddot{\theta}.$$

Multiplying the second equation by $\dot{\theta}$ and integrating, we obtain

$$\tfrac{1}{2}ml\dot{\theta}^2=mg\cos\theta+\text{constant}.$$

We could have written down this equation directly as the energy equation, since the tension in the string does no work. If the speed $v=l\dot{\theta}$ is equal to u when $\theta=0$, this becomes

$$v^2 = u^2 - 2lg(1 - \cos\theta),$$

and hence, from our first equation,

$$T = mu^2/l + mg(3\cos\theta - 2).$$

If the particle starts from $\theta = 0$ its swing will continue until either $v = 0$, which requires $\cos\theta = 1 - u^2/2lg$, or $T = 0$ (when the string becomes slack), which requires $\cos\theta = \frac{2}{3}(1 - u^2/2lg)$. Since initially $\cos\theta = 1$ the first of these possibilities will occur first if $u^2 < 2lg$, and the particle will come to rest with $\theta < \frac{1}{2}\pi$ and $T > 0$; it will then swing back along its previous path, performing periodic oscillations. If $2lg < u^2 < 5lg$ the second possibility will occur first, for a value of θ in the range $\frac{1}{2}\pi < \theta < \pi$ and with $v > 0$. The particle will then commence a free trajectory with the string slack. If $u^2 > 5lg$ neither T nor v ever vanishes (since $\cos\theta \geqslant -1$ for all θ) and so the particle performs a succession of vertical circles, without change of direction. There are thus three fundamentally different types of motion which may occur.

If u is very small, θ remains small thoughout the motion and the energy equation can be written approximately as

$$v^2 = l^2\dot{\theta}^2 = u^2 - lg\theta^2.$$

By comparison with Example 3 of section 2.3, this is the equation for simple harmonic motion of period $2\pi(l/g)^{\frac{1}{2}}$.

It is worth noting that exactly the same analysis applies if the particle moves on the inside of a smooth spherical bowl of centre O and radius l. At a smooth surface the reaction is perpendicular to the surface, and so acts towards O in this case. Tangential or *frictional* forces occur only if the surfaces in contact are rough.

2.7. MOTION UNDER A CENTRAL FORCE

The motion of a particle under the influence of a force directed towards a fixed point O, and of magnitude depending solely on the particle's distance from O, has many physical applications. In particular it enables us to discuss the orbits of planets and comets under the gravitational attraction of the sun.

Suppose that the force on a particle of mass m is $\mathbf{F}=mf(r)\hat{\mathbf{r}}$, so that the force is radial and of magnitude $f(r)$ per unit mass. The moment of this force about O is zero, and hence by (28) the angular momentum

$$\mathbf{H}=\mathbf{r}\wedge m\dot{\mathbf{r}}=\text{constant}. \tag{41}$$

Throughout the motion $\mathbf{r}$ and $\dot{\mathbf{r}}$ must be perpendicular to $\mathbf{H}$, and thus the motion lies entirely in the plane through O perpendicular to $\mathbf{H}$.

The potential energy per unit mass is $V(r)=-\int f(r)dr$, as in section 2.3, and hence the energy equation is

$$\tfrac{1}{2}\dot{\mathbf{r}}^2+V(r)=\text{constant}. \tag{42}$$

The lower limit in the integral for $V(r)$ may be chosen arbitrarily; it merely modifies the value of the constant in (42).

It is usually convenient to use polar co-ordinates in the plane of motion. Then $\mathbf{r}=r\mathbf{i}$, $\dot{\mathbf{r}}=\dot{r}\mathbf{i}+r\dot{\theta}\mathbf{j}$ as shown in (39) and $\mathbf{H}=mr^2\dot{\theta}\mathbf{k}$. Equations (41) and (42) become

$$r^2\dot{\theta}=h, \tag{43}$$

$$\tfrac{1}{2}\dot{r}^2+\tfrac{1}{2}r^2\dot{\theta}^2+V(r)=E, \tag{44}$$

where h and E are constants, being respectively the angular momentum per unit mass about the normal at O to the plane of motion, and the total energy per unit mass. We may use (43) to write (44) as

$$\tfrac{1}{2}\dot{r}^2+\tfrac{1}{2}h^2/r^2+V(r)=E. \tag{45}$$

In principle the motion is now fully determined. Equation (45) enables us to write dt/dr as a function of r, and we may integrate to find t as a function of r. Then (43) may be solved to obtain θ as a function of r or t. Alternatively we may calculate the orbit directly, using the equation

$$\left(\frac{dr}{d\theta}\right)^2=\frac{\dot{r}^2}{\dot{\theta}^2}=2\frac{r^4}{h^2}\left\{E-\frac{1}{2}\frac{h^2}{r^2}-V(r)\right\}. \tag{46}$$

The ease with which the actual integrations can be carried out depends, of course, on the particular form of $f(r)$.

In the case of gravity, *Newton's law of gravitation* states that the force of attraction between two particles of masses m and m' at a distance r apart has magnitude $\gamma mm'/r^2$, where γ is the universal gravitational constant. As will be shown in Example 4 of section 3.1, if m' is much greater than m we may treat m' as fixed at O and write $\mu=\gamma m'$, so that

$$f(r)=-\mu/r^2, \quad V(r)=-\mu/r.$$

The integration of equation (46) is facilitated by the substitution $r=1/u$. The equation becomes

$$\left(\frac{du}{d\theta}\right)^2=\frac{2E}{h^2}-u^2+\frac{2\mu u}{h^2}=\frac{2E}{h^2}+\frac{\mu^2}{h^4}-\left(u-\frac{\mu}{h^2}\right)^2,$$

and on taking the square root and integrating the resulting equation for $d\theta/du$ we obtain

$$\theta=\sin^{-1}\left\{\frac{u-\mu/h^2}{\left(\dfrac{2E}{h^2}+\dfrac{\mu^2}{h^4}\right)^{\frac{1}{2}}}\right\}+\tfrac{1}{2}\pi,$$

choosing the arbitrary constant in θ conveniently, or

$$l/r=1-e\cos\theta \tag{47}$$

where

$$l=h^2/\mu, \quad e^2=1+2Eh^2/\mu^2. \tag{48}$$

Equation (47) is the standard form in polar co-ordinates of a conic with semi-latus rectum l, eccentricity e, and the origin as one focus. The conic is an ellipse for $e<1$, a parabola for $e=1$, and a hyperbola for $e>1$. From (48) these three cases correspond to $E<0$, $E=0$ and $E>0$ respectively. The energy equation (44) may be written as

$$\tfrac{1}{2}v^2-\mu/r=E, \tag{49}$$

so the three cases also correspond to $v^2<,=$ or $>2\mu/r$. The nature of the orbit thus depends solely on the speed at a given position, and not on the direction of motion. For an ellipse the semi-major axis a is given by $l=a(1-e^2)$, and use of (48) shows that equation (49) may be written as $v^2=\mu(2/r-1/a)$. When $r=a$ (constant), $v^2=\mu/a$, confirming that for motion in a circle the centrifugal force mv^2/a balances the gravitational force $\mu m/a^2$. For a hyperbola $l=a(e^2-1)$, and hence $v^2=\mu(2/r+1/a)$. These forms are often convenient.

Example 1. *Show that* h *is equal to twice the rate at which the radius vector sweeps out area.*

In small time δt, with changes δr and $\delta\theta$ in r and θ, the area of the triangle formed by the initial and final radii is approximately $\frac{1}{2}r\,.\,r\delta\theta$. The rate at which area is swept out is therefore $\frac{1}{2}r^2\dot{\theta}=\frac{1}{2}h$.

Example 2. *A particle moves under a central attractive force varying inversely as the square of the distance. The particle approaches from a large distance with speed* u, *and if undeflected would pass at a distance* d *from the centre of force. Calculate the angle between the particle's initial and final directions of motion.*

At great distance $\mathbf{H}=\mathbf{r}\wedge m\dot{\mathbf{r}}=mdu\mathbf{k}$, and hence $h=du$. Also $E=\frac{1}{2}u^2$, since $V=0$ at infinity. Equation (48) now gives $e^2=1+u^4d^2/\mu^2$. From (47), when $r=\infty$, $\sec\theta=e$ and hence $\tan\theta=(e^2-1)^{\frac{1}{2}}=\pm u^2d/\mu$.

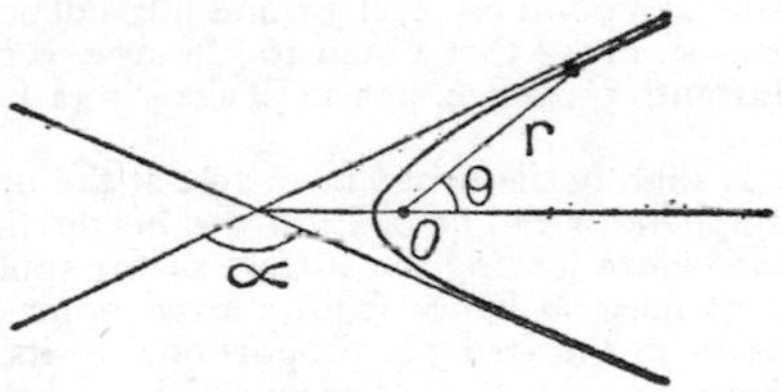

Figure 18

From Fig. 18, the total deflection is

$$\alpha=\pi-2\tan^{-1}(u^2d/\mu)=2\tan^{-1}(\mu/u^2d).$$

If it is known that a particle is moving under a central force a knowledge of its path enables us to calculate the law of force, since $dr/d\theta$ can be expressed as a function of r and then $V(r)$ found from (46). Differentiation with respect to r then gives $f(r)$, depending only on the parameter h.

Example 3. *Find the law of force if the path is log* $r=k\theta$.

Here $dr/d\theta=kr$, and (46) gives $V(r)=E-\frac{1}{2}h^2/r^2-\frac{1}{2}k^2h^2/r^2$. Differentiating, since $dV/dr=-f(r)$, we have

$$f(r)=-(1+k^2)h^2/r^3.$$

Thus the motion is governed by an inverse cube law of force, directed towards O.

EXERCISES ON CHAPTER TWO

1. A body falling freely under gravity takes $\frac{1}{4}$ sec. to fall past a window of height 9 ft. If the body started from rest, find the height from which it fell. [Take $g=32$ ft./sec.^{2}.]

2. A ball is thrown vertically upwards with speed U. If air resistance is proportional to the square of the speed and the terminal speed is V, find the speed of the ball when it returns to its starting point.

3. Find the angular momentum about O of a particle of mass 3 at the point $\mathbf{i}-4\mathbf{j}+6\mathbf{k}$ moving with velocity $3\mathbf{i}-2\mathbf{k}$. Find also the angular momentum about the line OP, where P is the point $2\mathbf{i}-\mathbf{j}-2\mathbf{k}$.

4. Calculate the maximum range of a projectile fired with speed u up a plane inclined at an angle β to the horizontal, and the elevation of the gun above the horizontal in this case, ignoring air resistance.

5. An explosion at a point on level ground hurls debris in all directions with speed 80 ft./sec. Prove that a man 100 ft. away is in danger of being struck at two instants $5/\sqrt{2}$ sec. apart. [Take $g=32$ ft./sec.^{2} and ignore air resistance.]

6. A particle is slightly disturbed from rest at the uppermost point of a smooth fixed sphere of radius b. Find the height through which the particle descends before leaving the surface of the sphere.

7. A particle of mass m hangs from a fixed point by a light elastic string. The tension in the string is proportional to its extension, which in the equilibrium position is l. If the particle makes oscillations in a vertical line, find its greatest speed if the string is not to become slack.

8. An aeroplane of mass 100 tons flies north along a parallel of longitude at 500 m.p.h. Calculate the transverse force needed to keep the aeroplane on course when in latitude 50°N.

9. A particle P of mass m moves under the action of a force of magnitude mkr directed towards a fixed point O, where r is the distance OP.

If the particle is projected at $r=b$ with speed V perpendicular to OP, show that it describes an orbit between $r=b$ and $r=V/\sqrt{k}$.

10. A particle moves in an ellipse under a central attractive force varying inversely as the square of the distance. If the particle's greatest speed is n times its least speed, find the eccentricity of the orbit.

CHAPTER THREE

Dynamics of a System

3.1. MOMENTUM

Our definition of a particle was formulated with special regard to the possibility of treating larger bodies as assemblages or systems of particles. We must now consider how dynamical laws for a system may be deduced from the laws for a single particle. The extra complication is that each particle is subject to internal forces of interaction with other particles of the system in addition to any externally applied force, and usually we neither know nor particularly wish to know the magnitudes of these internal forces. It is therefore vital to see how much information about the motion we can obtain without any detailed knowledge of them.

Let the typical particle of the system have mass m_i and be at the point with position vector $\mathbf{r}_i$. For a system of N particles, i takes the values 1, 2, . . ., N. Newton's second law of motion (26) for the ith particle can be written in the form

$$m_i\ddot{\mathbf{r}}_i = \mathbf{F}_i^* = \mathbf{F}_i + \sum_j \mathbf{F}_{ij}. \qquad (50)$$

Here $\mathbf{F}_i^*$ is the total force on the ith particle, $\mathbf{F}_i$ is the external force, and $\mathbf{F}_{ij}$ is the force due to the jth particle. The summation is over all values of j from 1 to N, and $\mathbf{F}_{ii}=0$ since a particle can have no influence on itself. Newton's third law, which we have not used previously, states that

$$\mathbf{F}_{ji} = -\mathbf{F}_{ij} \qquad (51)$$

for all values of i and j. If we add together the set of equations (50) corresponding to all values of i from 1 to N, the internal forces therefore all cancel in pairs, and we have

$$\Sigma m_i \ddot{\mathbf{r}}_i = \Sigma \mathbf{F}_i^* = \Sigma \mathbf{F}_i, \qquad (52)$$

each summation running from $i=1$ to $i=N$.

We now define $\Sigma m_i \dot{\mathbf{r}}_i = \Sigma \mathbf{M}_i = \mathbf{M}$ as the momentum of the system, being the sum of the momenta of the individual particles. Then

$$\frac{d\mathbf{M}}{dt} = \Sigma m_i \ddot{\mathbf{r}}_i = \Sigma \mathbf{F}_i = \mathbf{F}. \qquad (53)$$

Thus the rate of change of momentum of the system is equal to $\mathbf{F}$, the sum of the external forces. To avoid possible confusion with angular momentum, $\mathbf{M}$ is often referred to as the *linear momentum* of the system. If $\mathbf{F}=0$, $\mathbf{M}=$constant, which is the principle of the conservation of momentum for a system.

Still further simplification is possible. We define the *centre of mass* $\bar{\mathbf{r}}$ by the equation

$$\Sigma m_i \mathbf{r}_i = (\Sigma m_i)\bar{\mathbf{r}} = m\bar{\mathbf{r}} \qquad (54)$$

where m is the total mass of the system. Differentiating with respect to the time, we have

$$\Sigma m_i \dot{\mathbf{r}}_i = \mathbf{M} = m\dot{\bar{\mathbf{r}}}, \qquad (55)$$

and so the momentum of the system is equal to that of the total mass, moving with velocity of the centre of mass. Equation (53) may be written

$$m\ddot{\bar{\mathbf{r}}} = \mathbf{F}. \qquad (56)$$

This supports the self-consistency of our concept of a particle as far as the momentum is concerned, for we see by comparison with (26) that the system behaves just like a single particle of mass m at $\bar{\mathbf{r}}$, acted on by the external forces applied to the system.

It must be emphasized that there is no restriction at all on the types of internal forces which may be present, since it is well confirmed that all forces met with in practice do obey (51). Frictional, viscous and explosive forces are all permitted. For example, if a cat is swung and released so that its parabolic path as a projectile would take it into a bath of water—an experiment which the reader will of course not attempt to reproduce—no contortions by the cat can prevent it from landing in the bath (though it may manage to do so feet first).

Example 1. *A rod of mass* kx *per unit length extends from* x=0 *to* x=l. *Calculate the position of its centre of mass. Deduce the momentum of the rod when it rotates with constant angular velocity* ω *about a vertical axis perpendicular to its length through the end* x=0, *and the horizontal force at the axis required to maintain the motion.*

For a continuous body the summations in our formulae become integrals. At x, the particle of length dx has mass $kx\,dx$. The total mass $m=\int_0^l kx\,dx=\frac{1}{2}kl^2$. From (54), $m\bar{x}=\int_0^l kx^2dx=\frac{1}{3}kl^3$. Hence $\bar{x}=\frac{2}{3}l$. The centre of mass has speed $\omega\bar{x}$, and the magnitude of the momentum is $M=m\omega\bar{x}=\frac{1}{3}kl^3\omega$. From (56) and (39) the force at the axis is radial and of magnitude $m\omega^2\bar{x}=\frac{1}{3}kl^3\omega^2$.

Example 2. *A rocket of mass* m(t) *emits mass backwards at a speed* u *relative to the rocket at a constant rate* k(=− dm/dt). *Ignoring gravity and air resistance, calculate the force on the rocket and its speed* v *at time* t, *if at time* t=0 *its speed is* v_0 *and its mass* m_0.

In time δt the rocket's speed increases by δv. Since there are no external forces, there is no change in the momentum of the material which made up the rocket at time t, and hence $(m-k\delta t)\delta v-uk\delta t=0$. Letting $\delta t\to 0$, we have $dv/dt=uk/m=uk/(m_0-kt)$. Integrating,

$$v=-u\log(m_0-kt)+\text{constant}=v_0+u\log(m_0/m).$$

The force required to accelerate the mass $m-k\delta t$ is $(m-k\delta t)\,dv/dt$. Letting $\delta t\to 0$, we see that

$$F=m\,dv/dt=uk.$$

Note that the rate of change of momentum of the rocket,

$$\frac{d}{dt}(mv)=m\frac{dv}{dt}+v\frac{dm}{dt}=(u-v)k,$$

is *not* equal to the force on the rocket, since the mass emitted carries momentum away with it.

Example 3. *A jet of water of cross-section* A, *density* ρ *and speed* v, *strikes a wall at right angles and spreads out smoothly over it. Calculate the force on the wall.*

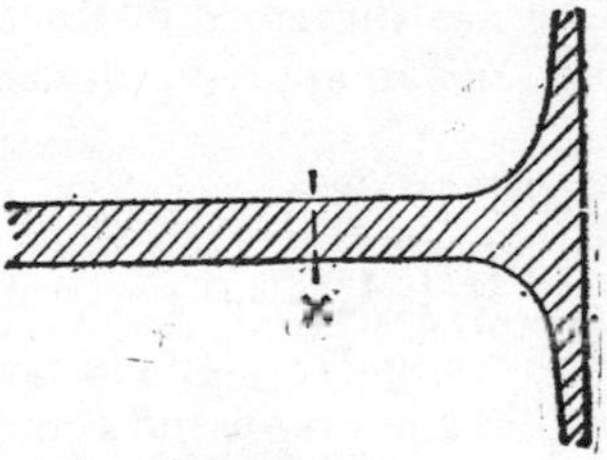

Figure 19

In time δt a volume $Av\delta t$ and hence a mass $\rho Av\delta t$ of fluid passes the point X in Fig. 19, carrying momentum $\rho Av^2\delta t$ perpendicular to the wall. There is no change in the total momentum in this direction between X and the wall, and so there must be a force $-\rho Av^2$ on the fluid to produce a counterbalancing loss of momentum. The reaction on the wall is therefore ρAv^2, parallel to the jet.

Example 4. *Two particles* A *and* A′ *of masses* m *and* m′ *move under their mutual gravitational attraction. Prove that the motion of* A *relative to* A′ *is the same as if* A′ *were fixed but had mass* m+m′.

The centre of mass X of the system may be considered as fixed in a Newtonian frame, since there are no external forces. Suppose $\overrightarrow{A'A}=\mathbf{r}$, $\overrightarrow{XA}=\mathbf{s}$, $\overrightarrow{A'X}=\mathbf{s}'$. Then $\mathbf{r}=\mathbf{s}'+\mathbf{s}$, $m\mathbf{s}=m'\mathbf{s}'$, and hence $(m+m')\mathbf{s}=m'\mathbf{r}$. The particle A is subject to the force $-(\gamma mm'/r^2)\hat{\mathbf{r}}$ and its equation of motion is

$$m\ddot{\mathbf{s}} = -\frac{\gamma mm'}{r^2}\hat{\mathbf{r}}, \text{ or } \ddot{\mathbf{r}} = -\frac{\gamma(m+m')}{r^2}\hat{\mathbf{r}}.$$

This is precisely the equation for motion under the attraction of a fixed mass $m+m'$ at the origin. When m is much smaller than m', as for the motion of the earth round the sun or of the moon round the earth, the difference between m' and $m+m'$ is not important and also the distance $A'X$ is relatively small. We are therefore justified in treating the larger mass as fixed, and in applying the analysis of section 2.7.

3.2. ANGULAR MOMENTUM

We now study the angular momentum of a system of particles. Suppose that the origin and centre of moments O is fixed in a Newtonian frame of reference. For the ith particle, take the vector product with $\mathbf{r}_i$ of the equation of motion (50). Then

$$\mathbf{r}_i \wedge m_i\ddot{\mathbf{r}}_i = \mathbf{r}_i \wedge \mathbf{F}_i^* = \mathbf{r}_i \wedge \mathbf{F}_i + \mathbf{r}_i \wedge \sum_j \mathbf{F}_{ij}. \quad (57)$$

Add the equations (57) for $i=1, 2, \ldots, N$ corresponding to the N particles of the system. The contribution from the interactions between the pth and q^{th} particles is

$$\mathbf{r}_p \wedge \mathbf{F}_{pq} + \mathbf{r}_q \wedge \mathbf{F}_{qp} = (\mathbf{r}_p - \mathbf{r}_q) \wedge \mathbf{F}_{pq} \quad (58)$$

by (51), Newton's third law. From the law as we have stated it, there is no reason to suppose that this expression vanishes. It is only zero either if $\mathbf{r}_p = \mathbf{r}_q$, or if $\mathbf{r}_p - \mathbf{r}_q$ is parallel to $\mathbf{F}_{pq}$, i.e. if $\mathbf{F}_{pq}$ acts along the line joining the particles. One of these conditions is satisfied for all the usual types of internal force. Gravitational, electrostatic and elastic forces act along the lines joining the particles, and for pressure and frictional forces between particles in contact we may consider $\mathbf{r}_p$ to be equal to $\mathbf{r}_q$. We shall assume here that all internal forces are such as to give no net contribution to the sum of equation (57). If at any time a type of interaction is found which does not obey this, the angular momentum laws which we now derive will not be applicable.

Define **H**, the angular momentum of the system about O, to be the sum of the angular momenta of the individual particles. Then

$$\mathbf{H} = \Sigma\, \mathbf{H}_i = \Sigma\, \mathbf{r}_i \wedge m_i \dot{\mathbf{r}}_i,$$

$$\frac{d\mathbf{H}}{dt} = \Sigma\, \mathbf{r}_i \wedge m_i \ddot{\mathbf{r}}_i + \Sigma\, \dot{\mathbf{r}}_i \wedge m_i \dot{\mathbf{r}}_i,$$

and each of the terms in the last sum is zero. The summation of equations (57) hence leads to

$$\frac{d\mathbf{H}}{dt} = \Sigma\, \mathbf{r}_i \wedge \mathbf{F}_i = \mathbf{G}, \tag{59}$$

where **G** is the sum of the moments about O of the external forces. Neither this equation for the rate of change of angular momentum nor equation (53) for the linear momentum requires any knowledge of the internal forces; it is for this reason that momentum and angular momentum are useful concepts for the discussion of the motion of extended bodies. If $\mathbf{G}=0$, $\mathbf{H}=$constant, which is the principle of the conservation of angular momentum for a system.

The position vector of the ith particle may be written as

$$\mathbf{r}_i = \bar{\mathbf{r}} + \mathbf{s}_i, \tag{60}$$

so that $\mathbf{s}_i$ is the position vector relative to the centre of mass. Multiplying by m_i and summing over the N particles of the system, we obtain

$$\Sigma\, m_i \mathbf{r}_i = (\Sigma\, m_i)\bar{\mathbf{r}} + \Sigma\, m_i \mathbf{s}_i.$$

Hence, from equation (54), and by differentiating,

$$\Sigma\, m_i \mathbf{s}_i = 0, \quad \Sigma\, m_i \dot{\mathbf{s}}_i = 0, \quad \Sigma\, m_i \ddot{\mathbf{s}}_i = 0. \tag{61}$$

Now

$$\begin{aligned}\mathbf{H} = \Sigma\, \mathbf{r}_i \wedge m_i \dot{\mathbf{r}}_i &= \Sigma\, (\bar{\mathbf{r}} + \mathbf{s}_i) \wedge m_i(\dot{\bar{\mathbf{r}}} + \dot{\mathbf{s}}_i) \\ &= \bar{\mathbf{r}} \wedge (\Sigma\, m_i)\dot{\bar{\mathbf{r}}} + \Sigma\, \mathbf{s}_i \wedge m_i \dot{\mathbf{s}}_i + \bar{\mathbf{r}} \wedge (\Sigma\, m_i \dot{\mathbf{s}}_i) \\ &\quad + (\Sigma\, m_i \mathbf{s}_i) \wedge \dot{\bar{\mathbf{r}}}.\end{aligned}$$

Here the constant factors have been taken outside the summations, and where convenient we have changed the position of m_i with relation to the vector product sign. From (61) the last two terms are zero, and the result may be written as

$$\mathbf{H}=\bar{\mathbf{r}} \wedge m\dot{\bar{\mathbf{r}}}+\overline{\mathbf{H}}, \tag{62}$$

where $\overline{\mathbf{H}}=\Sigma\, \mathbf{s}_i \wedge m_i\dot{\mathbf{s}}_i$. Thus the angular momentum of a system is equal to the angular momentum of the whole mass moving at the centre of mass, plus the angular momentum about the centre of mass of the motion of the system relative to the centre of mass. When calculating $\overline{\mathbf{H}}$ we may, if we wish, use the velocities of the particles relative to the origin O rather than relative to the centre of mass, since by (61)

$$\Sigma \mathbf{s}_i \wedge m_i\dot{\mathbf{r}}_i=\Sigma \mathbf{s}_i \wedge m_i(\dot{\bar{\mathbf{r}}}+\dot{\mathbf{s}}_i)=(\Sigma m_i\mathbf{s}_i) \wedge \dot{\bar{\mathbf{r}}}+\Sigma \mathbf{s}_i \wedge m_i\dot{\mathbf{s}}_i=\overline{\mathbf{H}}.$$

Angular momentum is particularly suitable for dealing with a rigid body. As shown in section 1.7, the motion of a rigid body may be completely specified by two vectors, the velocity $\mathbf{v}$ of a given point of the body and the angular velocity $\boldsymbol{\omega}$. Equations (53) and (59) for the momentum and the angular momentum provide two vector equations, and are sufficient to determine the whole motion. These equations show the importance of equivalent systems of forces and couples, discussed in section 1.8, for if a number of forces are applied to a rigid body, it is the resultant force and couple which completely determine the motion.

If $\mathbf{F}=\mathbf{G}=0$, (53) and (59) are satisfied if $\mathbf{v}=\boldsymbol{\omega}=0$, since then $\mathbf{M}=\mathbf{H}=0$. Also if $\mathbf{v}=\boldsymbol{\omega}=0$ for all time, $\mathbf{M}=\mathbf{H}=0$, and hence $\mathbf{F}=\mathbf{G}=0$. Thus $\mathbf{F}=\mathbf{G}=0$ is a necessary and sufficient condition for equilibrium to be possible for a rigid body.

The analysis of the general motion of three-dimensional bodies involves complications which we shall not be able to embark on here, but if the body and its motion effectively lie in a plane the problem is relatively straightforward. Consider a body in the xy-plane (a *lamina*), rotating about a fixed point O with angular velocity $\boldsymbol{\omega}=\omega\mathbf{k}$, where $\mathbf{k}$ is a unit vector in the z-direction. The element of mass m_i with position vector $\mathbf{r}_i$ has velocity of magnitude ωr_i perpendicular to $\mathbf{r}_i$, and hence the angular momentum of the body about O is

$$\mathbf{H}=\Sigma\, m_i\omega r_i^2\mathbf{k}=I\omega\mathbf{k},$$

where $I=\Sigma\, m_i\mathbf{r}_i^2$ is called the *moment of inertia* of the lamina about O, and depends only on its geometrical properties. More precisely, I is the moment of inertia about an axis through O in the z-direction.

The calculation of moments of inertia, like the calculation of centres of mass, is much studied in books on the integral calculus. For continuous bodies the summation in the definition of I becomes an integral. A useful result may be deduced from (62) when the body rotates about O as considered above. The equation gives

$$I\omega\mathbf{k}=m\omega\bar{r}^2\mathbf{k}+\bar{I}\omega\mathbf{k},$$

where $\bar{I}=\Sigma\, m_i\mathbf{s}_i^2$ is the moment of inertia about the mass centre, and therefore

$$I=\bar{I}+m\bar{r}^2. \tag{63}$$

This is known as the *theorem of parallel axes*. It states that the moment inertia about any axis is equal to the moment of inertia about a parallel axis through the centre of mass, plus the total mass multiplied by the square of the distance between the two axes. The theorem can be proved directly from equations (60) and (61), as the reader may verify. It shows that the moment of inertia about any point of a lamina is known once $\bar{I}$ and the position of the centre of mass have been determined.

Since moment of inertia has the dimensions of (mass) $\times$ (length)^2, it is sometimes convenient to write $I=mk^2$, where k is the *radius of gyration*. Equation (63) becomes

$$k^2=\bar{k}^2+\bar{r}^2.$$

The motion of a lamina rotating about O in the xy-plane is given by the z-component of the angular momentum equation (59). If θ is the angle which a given line in the lamina makes with a fixed direction in the plane, $\omega=\dot{\theta}$ and (59) gives

$$I\dot{\omega}=I\ddot{\theta}=G_z. \tag{64}$$

For a three-dimensional rigid body rotating with angular velocity $\omega\mathbf{k}$ about a fixed axis, the z-axis, the same analysis applies provided that in the definition of I, r_i is replaced by $\rho_i=(x_i^2+y_i^2)^{\frac{1}{2}}$, the distance of the mass m_i from the axis of rotation. The reader may verify that $H_z=I\omega$, and so (64) again applies. For this type of motion the z-components have no effect and so may be ignored, reducing the body once more to a lamina in the xy-plane.

Example 1. *Show that a constant force* **c** *per unit mass on a system of particles is equivalent to a force* m**c** *through the centre of mass.*

The total force is $\mathbf{F}=\Sigma\, m_i\mathbf{c}=(\Sigma\, m_i)\mathbf{c}=m\mathbf{c}$.

The sum of the moments of the forces about the centre of mass is

$$\mathbf{G}=\Sigma\, \mathbf{s}_i \wedge m_i\mathbf{c}=(\Sigma\, m_i\mathbf{s}_i)\wedge \mathbf{c}=0,$$

by (61). This proves the result, which enables us to identify the centre of mass of a body with its centre of gravity.

Example 2. *Calculate the moments of inertia about their centres of a uniform rod of mass* m *and length* 2a, *a uniform circular disc of mass* m *and radius* a, *and a uniform rectangular plate of mass* m *and sides* 2a, 2b.

In each case integrals replace the sums in the formulae. For the rod, the mass per unit length is $m/2a$, and so the element of mass is $(m/2a)dx$ and

$$I=\int_{-a}^{a}\frac{m}{2a}x^2dx=\tfrac{1}{3}ma^2.$$

For the disc, the elementary area $2\pi r\,dr$ between the radii r and $r+dr$ has mass $(m/\pi a^2)\,.\,2\pi r\,dr=(2mr/a^2)dr$, and

$$I=\int_{0}^{a}\frac{2mr}{a^2}r^2dr=\tfrac{1}{2}ma^2.$$

For the plate, divide the area into strips of length $2a$ and breadth dx, with centres at a distance x from O, the centre of the plate. The moment of inertia of such a strip about its centre is $\frac{1}{3}(mdx/2b)a^2$, and about O is $(mdx/2b)(\frac{1}{3}a^2+x^2)$, by (63). Hence

$$I=\int_{-b}^{b}\frac{m}{2b}(\tfrac{1}{3}a^2+x^2)dx=\tfrac{1}{3}m(a^2+b^2).$$

Example 3. *A skater of mass* 152 *lb. is rotating about a vertical axis with his arms outstretched at* 100 *revolutions per minute on smooth ice. He then folds his arms. If the radius of gyration of his arms (each of mass* 6 *lb.) is* 20 *in. when extended and* 5 *in. when folded, and if the remainder of his body has radius of gyration* 4 *in., calculate his new rate of rotation.*

Since there are no external moments about the axis, the angular momentum is unchanged. The skater's original moment of inertia was $140\times4^2+12\times20^2=2240+4800=7040$ lb. in.^{2}, and his new moment of inertia is $140\times4^2+12\times5^2=2240+300=2540$ lb. in.^{2}. Hence his new rate of rotation is $100\times7040\div2540=277$ r.p.m.

So far we have assumed that the origin and centre of moments O is fixed in a Newtonian frame of reference. We should like to be able to take moments about a moving point also, for two distinct reasons. First, as discussed in section 2.4, we have no reliable means of knowing what points may be considered to be at rest; and second, there are sometimes advantages in using a moving centre of moments even when the frame of reference is not in question, as for a wheel rolling down a slope for which it is convenient to take moments about the point of contact of the wheel with the slope.

Consider the second case first. Suppose that we have a moving origin O', so that $\overrightarrow{OO'}=\mathbf{a}$ as in Fig. 13. The *angular momentum about the moving origin* O' is

$$\mathbf{H}' = \Sigma\, \mathbf{r}_i' \wedge m_i\dot{\mathbf{r}}_i = \Sigma\,(\mathbf{r}_i - \mathbf{a}) \wedge m_i\dot{\mathbf{r}}_i = \mathbf{H} - \mathbf{a} \wedge \mathbf{M} \qquad (65)$$

and

$$\frac{d\mathbf{H}'}{dt} = \frac{d\mathbf{H}}{dt} - \mathbf{a} \wedge \frac{d\mathbf{M}}{dt} - \dot{\mathbf{a}} \wedge \mathbf{M}$$
$$= \Sigma\, \mathbf{r}_i \wedge \mathbf{F}_i - \mathbf{a} \wedge \Sigma\, \mathbf{F}_i - \dot{\mathbf{a}} \wedge \mathbf{M}$$

by equations (53) and (59). Note that there is no need to reconsider the effects of the internal forces. The sum of the moments of the external forces about O' is

$$\mathbf{G}' = \Sigma\, \mathbf{r}_i' \wedge \mathbf{F}_i = \Sigma\,(\mathbf{r}_i - \mathbf{a}) \wedge \mathbf{F}_i = \Sigma\, \mathbf{r}_i \wedge \mathbf{F}_i - \mathbf{a} \wedge \Sigma\, \mathbf{F}_i,$$

and hence

$$\frac{d\mathbf{H}'}{dt} = \mathbf{G}' - \dot{\mathbf{a}} \wedge \mathbf{M} = \mathbf{G}' - \dot{\mathbf{a}} \wedge m\dot{\bar{\mathbf{r}}}. \qquad (66)$$

This equation only reduces to the previous form (59) in special cases. It does so if $\dot{\mathbf{a}}=0$, so that O' is fixed relative to O, and also if $\dot{\mathbf{a}}$ is parallel to $\dot{\bar{\mathbf{r}}}$, which is true in particular if $\mathbf{a}=\bar{\mathbf{r}}$, so that O' is the centre of mass. This is an important case and leads to

$$\frac{d\bar{\mathbf{H}}}{dt} = \bar{\mathbf{G}} \qquad (67)$$

where $\bar{\mathbf{G}} = \Sigma\, \mathbf{s}_i \wedge \mathbf{F}_i$ is the sum of the moments of the external forces about the centre of mass. The fact that in general the rate of change of angular momentum about a moving origin is not equal to the sum of the moments of the applied forces will be illustrated in Example 1 of section 3.3.

In (65) the velocity used to form the angular momentum was the velocity relative to the fixed origin O. We can also define the *angular momentum relative to the moving origin O'* as

$$\mathbf{H}'' = \Sigma\, \mathbf{r}_i' \wedge m_i\dot{\mathbf{r}}_i', \qquad (68)$$

which involves the velocity $\dot{\mathbf{r}}_i'$ relative to O'. Since all quantities are now measured in the moving frame, the total dynamical effect is an inertial force $-m\ddot{\mathbf{a}}$ through the centre of mass, as is seen from Example 1 and from section 2.4, and the angular momentum equation is

$$\frac{d\mathbf{H}''}{dt} = \mathbf{G}' - \bar{\mathbf{r}}' \wedge m\ddot{\mathbf{a}}. \tag{69}$$

Again there is an extra term which vanishes in special cases only, the most important of which is where $\bar{\mathbf{r}}'=0$, i.e. O' is the centre of mass. The equation then reduces to (59) once more. The reader may establish equation (69) directly by expressing $\mathbf{H}''$ in terms of $\mathbf{H}$, as in the previous paragraph.

The practical moral to be drawn from this investigation of moving origins is that in dynamical problems it is safest to take as origin a fixed point, or the centre of mass.

Since $\bar{\mathbf{H}} = \Sigma\, \mathbf{s}_i \wedge m_i \dot{\mathbf{s}}_i$ and $\bar{\mathbf{G}} = \Sigma\, \mathbf{s}_i \wedge \mathbf{F}_i$, all the distances and velocities in equation (67) are measured from the centre of mass, and do not depend at all on the basic co-ordinate system. We have thus achieved the aim expressed in section 2.4, of finding a form of equation which does not involve the particular Newtonian frame of reference in which we stated the laws of motion. To sum up, the motion of the centre of mass of a system is given by (56) and the motion relative to the centre of mass is described by (67), neither equation requiring any knowledge of the internal forces on the system.

Example 4. *Two particles* A, B *of masses* m_A, m_B *and position vectors* $\mathbf{r}_A$, $\mathbf{r}_B$ *are free to move under their mutual gravitational attraction. Prove that the angular momentum of* B *relative to* A *is constant.*

The force on B acts towards A, so $\mathbf{G}'=0$, but A is neither the centre of mass of the system nor a fixed point, so the required result is not yet established. However, the acceleration of A is towards the centre of mass, so in (69) $\ddot{\mathbf{a}}$ is parallel to $\bar{\mathbf{r}}'$ and $\bar{\mathbf{r}}' \wedge m\ddot{\mathbf{a}}=0$. The result now follows.

We may conclude this discussion of angular momentum with a further glance at the unfortunate cat of page 57. Do our laws imply that, without the possibility of obtaining any reaction from the air, the cat could not turn so as to land feet first? Equation (67) shows that the cat's angular momentum about its centre of mass if initially zero must remain zero, and so, considered as a rigid body, the cat cannot acquire angular velocity. However, rotation is still possible by means of suitable flexing movements. An example more easily visualized is a man standing on a sheet of perfectly smooth ice who wishes to turn round. Starting with his arms close to his side he extends them both to his right. Then he swings them round horizontally in a half-circle until they are extended to his left. During this motion his body turns to the right, to keep his angular momentum zero. Then he drops his arms again. Moving his arms close to his body from his left side to his right causes only a slight rotation of his body to the left, since the moment of inertia of his arms is now much reduced. Thus he returns to his starting position facing more to the right, still with no angular velocity.

Finally we may use equations (50), (56) and (60) to write the equation of motion of the ith particle, relative to the centre of mass, in the form

$$m_i\ddot{\mathbf{s}}_i = m_i\ddot{\mathbf{r}}_i - m_i\ddot{\bar{\mathbf{r}}} = \mathbf{F}_i^* - (m_i/m)\,\mathbf{F}.$$

Like (67), this is entirely independent of the translational motion (though not rotation) of the particular frame of reference in which displacements are measured.

3.3. ENERGY

The kinetic energy of a system is defined to be the sum of the kinetic energies of the individual particles.

$$T = \tfrac{1}{2}\,\Sigma\, m_i\dot{\mathbf{r}}_i^2. \tag{70}$$

As for angular momentum in equation (62), there are often advantages in introducing the centre of mass $\bar{\mathbf{r}}$. Then

$$\begin{aligned} T &= \tfrac{1}{2}\,\Sigma m_i(\dot{\bar{\mathbf{r}}} + \dot{\mathbf{s}}_i)^2 = \tfrac{1}{2}(\Sigma m_i)\dot{\bar{\mathbf{r}}}^2 + \dot{\bar{\mathbf{r}}}\cdot(\Sigma m_i\dot{\mathbf{s}}_i) + \tfrac{1}{2}\,\Sigma m_i\dot{\mathbf{s}}_i^2 \\ &= \tfrac{1}{2}m\dot{\bar{\mathbf{r}}}^2 + \bar{T}, \end{aligned} \tag{71}$$

where $\bar{T} = \tfrac{1}{2}\,\Sigma m_i\dot{\mathbf{s}}_i^2$, from (61). This shows that the kinetic energy of a system of particles is equal to the kinetic energy of the total mass moving at the centre of mass, plus the kinetic energy of the particles' motion relative to the centre of mass.

For a rigid lamina rotating in its own plane about a point O, as considered in section 3.2,

$$T=\tfrac{1}{2}\Sigma\, m_i(\omega r_i)^2=\tfrac{1}{2}I\omega^2.$$

Once again the moment of inertia I is involved.

From (50) the energy equation for the ith particle is

$$\frac{d}{dt}(\tfrac{1}{2}m_i\dot{\mathbf{r}}_i^2)=\mathbf{F}_i^* \cdot \dot{\mathbf{r}}_i=\mathbf{F}_i \cdot \dot{\mathbf{r}}_i+\sum_j \mathbf{F}_{ij} \cdot \dot{\mathbf{r}}_i. \qquad (72)$$

When we sum the N equations (72) corresponding to the N particles of the system, the contribution from the interaction between the pth and q^{th} particles is $\mathbf{F}_{pq} \cdot (\dot{\mathbf{r}}_p-\dot{\mathbf{r}}_q)$, on making use of (51). This is the magnitude of the force multiplied by the component of the relative velocity of the two particles in the direction of the force, and is equal to the rate of working of the interaction.

In general, internal work is done in the deformation of an elastic body, in the motion of a system of charged or gravitating particles, and by frictional forces. But there is a wide class of interactions which do no work, including the reaction at a smooth surface on a sliding body, the reaction between bodies which roll on one another, the tension in an inextensible string, and the reaction at a smooth joint or pivot. In particular the interaction between any two particles of a rigid body does no work when, as we have assumed, it acts along the line joining the particles, for then $\mathbf{F}_{pq}=\lambda(\mathbf{r}_p-\mathbf{r}_q)$ and

$$\mathbf{F}_{pq} \cdot (\dot{\mathbf{r}}_p-\dot{\mathbf{r}}_q)=\tfrac{1}{2}\lambda\frac{d}{dt}(\mathbf{r}_p-\mathbf{r}_q)^2=\tfrac{1}{2}\lambda\frac{d}{dt}|\mathbf{r}_p-\mathbf{r}_q|^2$$

by equation (5), and this is zero since $|\mathbf{r}_p-\mathbf{r}_q|$ is constant.

If all the internal interactions are workless, the energy equation becomes

$$\frac{dT}{dt}=\Sigma\,\mathbf{F}_i \cdot \dot{\mathbf{r}}_i, \qquad (73)$$

and when the external forces are conservative
$\Sigma \mathbf{F}_i \cdot \dot{\mathbf{r}}_i = dV/dt$, and hence

$$T+V=\text{constant}, \tag{74}$$

where V is the potential energy.

Equations (71) and (62) give further support to the consistency of our definition of a particle. They show that when a body is sufficiently small for $\bar{T}$ and $\bar{\mathbf{H}}$ to be negligible compared with the kinetic energy and angular momentum of the overall motion, the body may be treated as a single particle of mass m at the point $\bar{\mathbf{r}}$. Furthermore, equations (73) and (59) show that only the external forces on the body affect the motion.

The following examples illustrate some of the results we have obtained. To improve his understanding of the principles of dynamics the reader should study worked examples with care, and should above all practise solving problems for himself.

Example 1. *A uniform disc of mass* m, *centre* C *and radius* a *has a particle of mass* m *fixed at a point* P *on its rim. The disc can roll in a vertical plane along a horizontal surface, and is slightly disturbed from rest with* P *vertically above* C. *Find the angular velocity of the disc when* P *is at the same height as* C.

Let PC make an angle θ with the upward vertical. The angular velocity of the disc is $\dot{\theta}$. If the horizontal velocity

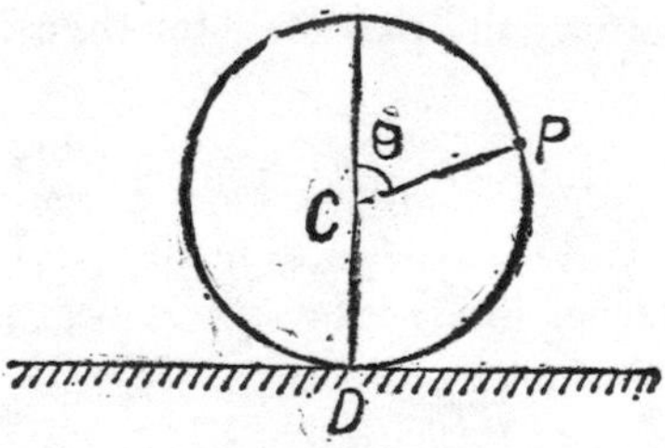

Figure 20

of C is u, the velocity of the point of contact D is $u - a\dot{\theta}$. This must be zero since there is no slipping; hence $u = a\dot{\theta}$ and the kinetic energy of the disc is $\frac{1}{2}ma^2\dot{\theta}^2 + \frac{1}{4}\ ma^2\dot{\theta}^2 = \frac{3}{4}ma^2\dot{\theta}^2$, by (71), since the disc's moment of inertia about its centre is $\frac{1}{2}ma^2$. The particle's velocity relative to C is $a\dot{\theta}$ perpendicular to CP, so the components of its total velocity are $a\dot{\theta}(1+\cos\theta)$ and $a\dot{\theta}\sin\theta$, and its kinetic energy is $\frac{1}{2}ma^2\dot{\theta}^2\{(1+\cos\theta)^2+\sin^2\theta\} = ma^2\dot{\theta}^2(1+\cos\theta)$. Thus for the system the kinetic energy is

$$T = ma^2\dot{\theta}^2(\tfrac{7}{4}+\cos\theta).$$

The gravitational potential energy measured from C is $V = mga\cos\theta$, and no other forces on the system do work. The energy equation is therefore

$$ma^2\dot{\theta}^2(\tfrac{7}{4}+\cos\theta)+mga\cos\theta = \text{constant} = mga$$

since when $\theta=0$, $\dot{\theta}=0$. When P is at the level of C, $\theta=\frac{1}{2}\pi$ and this equation shows that the disc's angular velocity is then $\dot{\theta}=(4g/7a)^{\frac{1}{2}}$.

An alternative method of finding T is to observe that the system is instantaneously rotating with angular velocity $\dot{\theta}$ about the disc's point of contact D. By (63), the disc's moment of inertia about D is $\frac{3}{2}ma^2$. Also $DP=2a\cos\frac{1}{2}\theta$. The same value for T is obtained.

Note that care would be needed in using angular momentum about D, since this point is neither fixed nor the centre of mass of the system. About D the angular momentum is $H=\frac{3}{2}ma^2\dot{\theta}+4ma^2\dot{\theta}\cos^2\frac{1}{2}\theta$, and the moment of the external forces is $G=mga\sin\theta$. The reader may verify that dH/dt is not equal to G, by comparing with the result of differentiating the energy equation found above. He should check that the discrepancy is as given by the last term of equation (66), which in this example has the value $-ma^2\dot{\theta}^2\sin\theta$.

Example 2. *A uniform rod* AB *of mass* m *and length* 2a *is balanced on its end* A *on a smooth horizontal table, and then slightly disturbed. Calculate the reaction* R *of the table when the rod makes an angle* θ *with the vertical.*

There are no horizontal forces on the rod, so its centre O moves in a vertical line, by (56). The height of O above the plane is $a\cos\theta$ and so its speed is $a\dot{\theta}\sin\theta$ downwards. Hence $T=\frac{1}{2}ma^2\dot{\theta}^2\sin^2\theta+\frac{1}{6}ma^2\dot{\theta}^2$, by (71). Also $V=mga\cos\theta$ since the reaction at A does no work, and the energy equation is

$$a\dot{\theta}^2(1+3\sin^2\theta)+6g\cos\theta=6g$$

since $\dot{\theta}=0$ when $\theta=0$. For the motion of the centre of mass O, (56) gives

$$mg-R=m(d/dt)(a\dot{\theta}\sin\theta)=ma\ddot{\theta}\sin\theta+ma\dot{\theta}^2\cos\theta.$$

Differentiating the energy equation and dividing by $2\dot{\theta}$, we have

$$a\ddot{\theta}(1+3\sin^2\theta)+3a\dot{\theta}^2\sin\theta\cos\theta-3g\sin\theta=0.$$

Using this equation and the energy equation to express $\ddot{\theta}$ and $\dot{\theta}$ in terms of θ, we obtain after some manipulation

$$R=\frac{4-6\cos\theta+3\cos^2\theta}{(1+3\sin^2\theta)^2}mg.$$

Example 3. *Discuss the motion of a compound pendulum, consisting of a lamina of mass* m *free to rotate in a vertical plane about a fixed point* O *at a distance* d *from the centre of mass* A, *if the moment of inertia of the lamina about* O *is* mk².

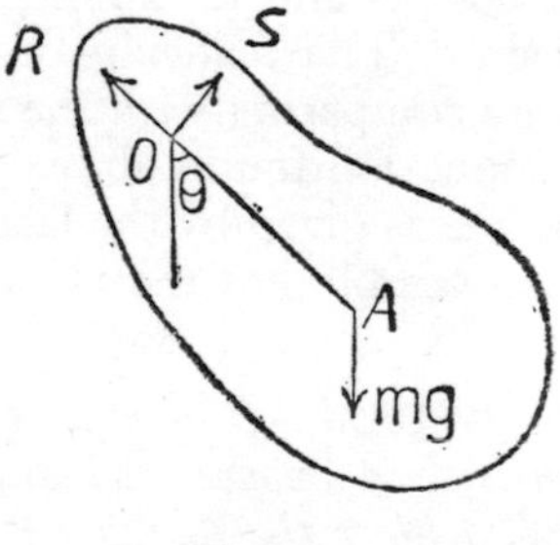

Figure 21

Let OA make an angle θ with the downward vertical, as shown in Fig. 21. The energy equation is

$$\tfrac{1}{2}mk^2\dot{\theta}^2 - mgd\cos\theta = \text{constant}$$

since the reaction at O does no work. This is identical with the energy equation for a simple pendulum of length $l = k^2/d$, as found in Example 1 of section 2.6. Thus small oscillations are simple harmonic with period $2\pi\sqrt{(k^2/dg)}$.

The length l of the equivalent simple pendulum varies with the position of the point of support O. By the theorem of parallel axes $k^2 = \bar{k}^2 + d^2$, and therefore the minimum value of l is $2\bar{k}$ when $d = \bar{k}$, as the reader may verify.

Since there is no restriction as to the direction of the reaction at O there are only two fundamental types of motion for a compound pendulum, depending on whether or not the angular velocity falls to zero before A is vertically above O. If the pendulum starts from $\theta = 0$ with angular velocity ω_0, the energy equation is

$$\tfrac{1}{2}mk^2\dot{\theta}^2 - mgd\cos\theta = \tfrac{1}{2}mk^2\omega_0^2 - mgd.$$

The pendulum comes to rest with $0 < \theta < \pi$ if $\omega_0^2 < 4gd/k^2$, and periodic oscillations then ensue. If $\omega_0^2 > 4gd/k^2$ the pendulum describes a succession of vertical circles.

The reaction on the lamina at O may be found by considering the momentum of the lamina. Let R, S be the components of the reaction parallel and perpendicular to OA, as shown in Fig. 21. Then from (39) and (56)

$$mg\cos\theta - R = -md\dot{\theta}^2, \quad S - mg\sin\theta = md\ddot{\theta}.$$

Differentiating the energy equation obtained above gives $mk\ddot{\theta} + mgd\sin\theta = 0$ (this could have been written down as the equation for angular momentum about O) and hence

$$R = mg\cos\theta + md\omega_0^2 - \frac{2mgd^2}{k^2}(1 - \cos\theta), \quad S = \frac{mg}{k^2}(k^2 - d^2)\sin\theta.$$

3.4. IMPULSIVE MOTION

When solid bodies collide, there are very large forces of interaction which last only for a very short time. These forces produce finite changes of velocity and momentum, but the changes in the position of the system in the short time interval can be ignored. The laws governing the dynamics of impacts can be deduced from the laws we have developed here.

Suppose that we integrate the momentum equation (53) for a system from $t=t_0$ to $t=t_1$, the duration of the impact. We obtain

$$\mathbf{I}=\Sigma\,\mathbf{I}_i=[\mathbf{M}]_{t_0}^{t_1}=\mathbf{M}_1-\mathbf{M}_0 \tag{75}$$

where $\mathbf{I}_i=\int_{t_0}^{t_1}\mathbf{F}_i(t)dt$ is the *impulse* of the exteınal force $\mathbf{F}_i(t)$ on the ith particle, $\mathbf{I}=\int_{t_0}^{t_1}\mathbf{F}(t)dt$, and the suffixes 0 and 1 denote values before and after the impact. By section 1.6, $\mathbf{I}_i$ and $\mathbf{I}$ are vectors. Equation (75) shows that the change of momentum of a system is equal to the sum of the external impulses. Note that there is no need to reconsider the internal forces on the system, since (53) applies at each instant.

If the force $\mathbf{F}_i$ is constant during the time interval, $\mathbf{I}_i=(t_1-t_0)\mathbf{F}_i$, but there is no theoretical reason for $\mathbf{F}_i$ to have constant magnitude or even constant direction. Since t_1-t_0 is small, ordinary forces such as gravity can produce no significant contribution to $\mathbf{I}_i$ and so can be ignored during the impact.

In the integration of the angular momentum equation (59), the positions $\mathbf{r}_i$ of the particles may be treated as constants. Hence

$$\int_{t_0}^{t_1}\mathbf{G}\,dt=\Sigma\left(\mathbf{r}_i\wedge\int_{t_0}^{t_1}\mathbf{F}_i dt\right)=\Sigma\,\mathbf{r}_i\wedge\mathbf{I}_i,$$

and we obtain

$$\mathbf{J}=\Sigma\,\mathbf{r}_i\wedge\mathbf{I}_i=[\mathbf{H}]_{t_0}^{t_1}=\mathbf{H}_1-\mathbf{H}_0, \tag{76}$$

where $\mathbf{J}$ is the sum of the moments of the external impulses.

As in section 3.2, this equation holds when the origin is fixed or is at the centre of mass of the system.

These laws are simpler to deal with than the equations for non-impulsive motion, since they are not differential equations but relate directly the velocities before and after the impact.

Example 1. *A uniform rod* AB *of length* 2a *and mass* m *is rotating about its centre with angular velocity* ω_0 *when the end* A *is suddenly fixed. Calculate the new angular velocity of the rod and the impulsive reaction at* A.

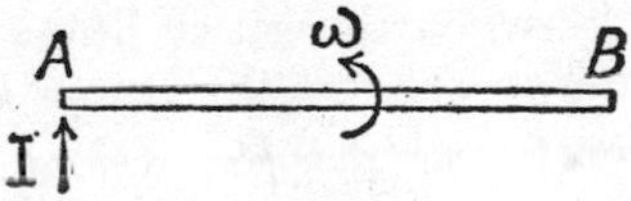

Figure 22

Let the impulse on the road at A be I, as shown in Fig. 22. If the rod has angular velocity ω_1 about A after the impact, its centre has speed $a\omega_1$ perpendicular to AB and (75) gives

$$I = ma\omega_1.$$

Relative to the centre of mass, (76) gives

$$-Ia = \tfrac{1}{3}ma^2\omega_1 - \tfrac{1}{3}ma^2\omega_0.$$

Solving these equations, we obtain $\omega_1 = \frac{1}{4}\omega_0$, $I = \frac{1}{4}ma\omega_0$.

In general, kinetic energy is lost in an impact. The reader may verify that in Example 1 only one-quarter of the kinetic energy remains. An instructive case to consider is the direct impact of two particles of masses m, m' moving along the x-axis with velocities v, v'. The kinetic energy of the motion relative to the centre of mass is

$$\overline{T} = \tfrac{1}{2}m(v - \bar{v})^2 + \tfrac{1}{2}m'(v' - \bar{v}')^2 = \tfrac{1}{2}\frac{mm'}{m + m'}(v - v')^2 \qquad (77)$$

as the reader may readily verify. In the collision this kinetic energy will be unchanged only if $v_1 - v_1' = \pm(v_0 - v_0')$. The positive sign implies no change in velocity of either particle and hence no impact. The negative sign implies that the relative velocity is reversed, with unchanged magnitude. This is known as a perfectly elastic collision.

An experimental law due to Newton, which is satisfactorily obeyed in collisions between hard bodies, is

$$v_1 - v_1' = -e(v_0 - v_0') \tag{78}$$

where the constant e is the *coefficient of restitution.* For a perfectly elastic collision $e=1$, and for a completely inelastic collision with no rebound $e=0$. For ordinary materials e lies between these two limits. It follows from (77) that the collision reduces the kinetic energy relative to the centre of mass by a factor e^2.

The kinetic energy associated with the motion of the centre of mass of the system is, of course, unaltered. Equation (75) indicates that the internal impulses caused by the impact can have no effect on the velocity of the centre of mass.

In collisions between smooth rigid bodies in three-dimensional motion, experiment shows that it is reasonable to assume that the velocity components in the direction of the normal at the point of contact satisfy (78), and that perpendicular velocity components are unaffected.

Example 2. *A smooth sphere of mass* m *strikes a second sphere of mass* 2m *at rest. After the collision their directions of motion are at right angles. Find* e.

Let the velocity components along and perpendicular to the line of impact be u, v for the mass m and u', v' for the mass $2m$. Then $v_1 = v_0$ and $v_1' = v_0' = 0$. Also by (78), $u_1 - u_1 = -eu_0$ and by (75), $mu_1 + 2mu_1' = mu_0$ since there are no external impulses. Solving these equations, we obtain $3u_1 = (1-2e)u_0$, $3u_1' = (1+e)u_0$. Since $v_1' = 0$, the velocities will be perpendicular if $u_1 = 0$, which requires $e = \frac{1}{2}$.

Example 3. *A uniform rod of mass* m *and length* 2a *is hanging freely from one end* O *when it is struck at a distance* b *from* O *by a bullet of mass* m′ *moving horizontally with speed* v. *The bullet becomes embedded in the rod, which swings through an angle* α. *Express* v *in terms of the other quantities.*

First consider the impact. Let ω be the angular velocity of the rod just after the bullet has become embedded. There is an impulsive reaction at O, but angular momentum about O is conserved and hence

$$m'vb=(\tfrac{4}{3}ma^2+m'b^2)\omega.$$

In the ensuing motion energy is conserved (although energy was lost in the impact) and hence

$$\tfrac{1}{2}(\tfrac{4}{3}ma^2+m'b^2)\omega^2=(ma+m'b)(1-\cos\alpha)g.$$

Eliminating ω, we obtain

$$m'^2b^2v^2=2g(\tfrac{4}{3}ma^2+m'b^2)(ma+m'b)(1-\cos\alpha),$$

which is the required equation.

This type of apparatus is known as a *ballistic pendulum*, and has been of much practical use in the development of artillery.

EXERCISES ON CHAPTER THREE

1. A man standing on a sheet of smooth ice sets himself in motion by throwing successively his two boots, each of mass m, in the same horizontal direction with speed v relative to himself. Find the man's final speed if his mass without his boots is m'.

2. A uniform circular disc of mass m, centre A and radius a is free to rotate in a vertical plane about a fixed point O, on its circumference. If the disc is disturbed from rest with A vertically above O, calculate the components of the reaction at O when OA is horizontal.

3. A truck of mass m moving with speed $3v$ collides with a second truck of mass $2m$ moving with speed v in the same direction. The two move on together. Calculate the impulsive reaction and the loss of kinetic energy in the collision.

4. The letter T is formed by rigidly joining together two uniform rods each of mass m and length $2a$. It is hung freely from the foot of the T, and allowed to make small oscillations under gravity in its own plane. Calculate the length of the equivalent simple pendulum.

5. Show that the moment of inertia of a uniform solid sphere of mass m and radius a about an axis through its centre is $\frac{2}{5}ma^2$. [Consider the sphere as a set of circular discs of thickness dz and radius $(a^2-z^2)^{\frac{1}{2}}$.]

6. A uniform sphere of mass m and radius a rolls down a rough plane inclined at an angle α to the horizontal. Show that the acceleration of the centre of the sphere is uniform and find its magnitude. Calculate the frictional force exerted on the sphere by the plane.

7. A uniform rod AB of length $2l$ and mass m is freely pivoted at O, where $OA=\frac{1}{2}l$. Find at what point X a blow perpendicular to the rod will cause no impulsive reaction at O. [X is known as the *centre of percussion*.]

8. Three particles A, B, C, each of mass m, lie in order on a straight line. B and C are initially at rest and A has speed U directly towards B. If the coefficient of restitution at each impact is 0·2, and there are no external forces, show that there will be just three impacts and find the final speed of each particle.

9. A uniform circular disc of mass $6m$ and radius a is free to turn in a horizontal plane about a fixed vertical axis through its centre. A mouse of mass m (which may be treated as a particle) is standing on the disc at its rim, the whole system being at rest. Suddenly the mouse starts to run round the rim with uniform speed v relative to the disc. Find the time after which the mouse returns to its starting point in space.

10. Find the time in the preceding question if the disc, instead of being pivoted, rests on a perfectly smooth table.

Answers to Exercises

Chapter I

1. $7; \frac{6}{7}, -\frac{2}{7}, \frac{3}{7}$. $2\sqrt{5}$; $-1/\sqrt{5}$, 0, $2/\sqrt{5}$. 12° 36′.
3. $\mathbf{r} \cdot (\mathbf{b} \wedge \mathbf{c}+\mathbf{c} \wedge \mathbf{a}+\mathbf{a} \wedge \mathbf{b})=[\mathbf{a}, \mathbf{b}, \mathbf{c}]$.
4. $5\mathbf{i}+4\mathbf{j}-3\mathbf{k}$, $5\mathbf{i}-4\mathbf{j}+3\mathbf{k}$.
6. $\mathbf{x}=\{(\mathbf{a} \cdot \mathbf{b})\mathbf{a}+\mathbf{b}+\mathbf{a} \wedge \mathbf{b}\}/(1+a^2)$.
7. 12.32 p.m., 8 sea miles.
8. $9\mathbf{i}-\mathbf{j}-32\mathbf{k}$, $-132/5$.
10. $-3\mathbf{i}+2\mathbf{j}+5\mathbf{k}$, $\mathbf{r}=-3\lambda\mathbf{i}+(2\lambda+10)\mathbf{j}+(5\lambda-3)\mathbf{k}$.

Chapter II

1. 16 ft. above the top of the window.
2. $\dfrac{VU}{\sqrt{(V^2+U^2)}}$.
3. $24\mathbf{i}+60\mathbf{j}+36\mathbf{k}$, -28.
4. $\dfrac{u^2}{g(1+\sin\beta)}, \dfrac{\pi}{4}+\dfrac{\beta}{2}$.
6. $\frac{1}{3}b$.
7. $\sqrt{(lg)}$.
8. 570 lb. wt. towards the west.
10. $(n-1)/(n+1)$.

Chapter III

1. $mv\left(\dfrac{1}{m'+m}+\dfrac{1}{m'+2m}\right)$.
2. $\frac{4}{3}mg$ horizontally, $\frac{1}{3}mg$ vertically upwards.
3. $\frac{4}{3}mv, \frac{4}{3}mv^2$.
4. $17a/9$.
6. $\frac{5}{7}g \sin\alpha$, $\frac{2}{7}mg \sin\alpha$ up the plane.
7. $AX=\frac{5}{8}l$.
8. $0{\cdot}304U$, $0{\cdot}336U$, $0{\cdot}36U$.
9. $8\pi a/3v$.
10. $18\pi a/7v$.

Index